The
ABCs
of
Environmental
Science

William B. Katz, PE

Government Institutes
Rockville, MD

Government Institutes, a Division of ABS Group Inc.
4 Research Place, Rockville, Maryland 20850, USA
Phone: (301) 921-2300
Fax: (301) 921-0373
Email: giinfo@govinst.com
Internet: http://www.govinst.com

Katz, William B., 1917-
     ABCs of environmental science / by William B. Katz.
        p. cm.
     Includes index.
     ISBN: 0-86587-627-4
     1.Environmental sciences. 2. Nature—Influence of human beings on. I. Title.
     GE105.K38   1998
     363.7—dc21   98-27546
                                                              CIP

Printed in the United States of America

Table of Contents

Preface

I am a prejudiced engineer! That is a contradiction. Engineers are taught to be objective, measuring things with instruments, correlating data, and making reasoned judgments that are impartial, accurate, and based on facts. I am also, among other things, a husband, father, grandfather, citizen, resident of my community, avid reader, and TV watcher. These different roles affect my emotions in various ways, as they do yours. Important (or unimportant) issues of the day tingle my consciousness. Some of the issues I know as much or as little about as any other person. On the other hand, I know a lot about some of the issues that capture headlines today. As an engineer, I work daily in close contact with the things and people involved with those issues. That is why I'm prejudiced.

You and I make decisions and ask our elected representatives to make decisions, often based more on emotions than on facts. Agencies that have to write and enforce regulations that control environmental matters are buffeted about by a host of concerned citizens and lawmakers. Some of these buffeters know a great deal about what they are trying to talk about. Some of them know very little. Many represent special-interest groups. That's OK. That is the way things work in our country, and I would not want it any other way. But the decisions that have to be made are so critical to our future (and by "our" I mean this planet Earth) that they simply must be based on the best facts available to us, not on emotions.

Contrary to what some motion pictures and printed words would lead us to believe, it is not true that all business is "bad." Businesses are run by people with children and grandchildren, people who like clean lakes to swim in, unpolluted rivers to fish in, and pure air to breathe, just like you and me. It is not true, as some business publications imply, that all special-interest groups are filled with "kooks" who are impractical, obstructionist, and uninformed. Many members of these groups are intelligent, educated, and have serious doubts about what is or is not being done to protect our future.

It is not true that all those in government are interested only in being reelected or preserving their jobs, and that they care not a whit about those who elected them. I have met too many legislators and "bureaucrats" who are truly concerned about doing what is right and best. It is not true that the "media" are interested only in headlines and care little about a balanced story or facts. I have read too many good stories while knowing the facts to believe that.

But I have also seen some truth in all the above generalizations about what is not good in government, the media, the special-interest groups, and business. So how do you know what to think in our increasingly technical world? That is where this prejudiced engineer enters the picture. In this book I will try to tell you, in simple, understandable language, what the environment is all about, how it works, and what impact we humans have on it. Any opinions I express will be my own, backed up as much as possible by information from others whose experience, judgment, and background I respect. Hopefully, when you are finished reading the book, you will have an understanding of how our planet is affected by everything we humans do. You will have enough information to discuss any environmental issue, to question anyone proposing solutions to environmental problems, and to start taking your own actions to save our earth from environmental disaster. Every small step counts! "He who lights one candle can light up the earth!"

About the Author

William Katz is a registered professional engineer in Illinois, Indiana, Michigan, and Wisconsin. He obtained a bachelor's degree from the University of Illinois and a master's degree from the Massachusetts Institute of Technology, both in chemical engineering. He is recently retired from an environmental business he founded over 25 years ago.

Mr. Katz is a published author on environmental subjects, including the chapter "Spill Prevention" in the McGraw-Hill *Handbook of Hazardous Materials*. He is an Emeritus (over 50 years) member and a Fellow of the American Institute of Chemical Engineers, a 50-year member of the American Chemical Society, a Fellow of the American Association for the Advancement of Science, and a former member of the National Environmental Training Association. He is a founding member and former vice-chairman of ASTM Committee F-20 on oil and hazardous material spill response and was elected an ASTM Fellow (Award of Merit) for those activities. He is a member of ASTM Committee E-50 on environmental assessment, a subcommittee of which writes standards for commercial real estate environmental audits. He is also a member of the National Society of Environmental Consultants, an organization of real estate environmental auditing professionals.

Mr. Katz was a founding member and has been elected an honorary director for life of the Spill Control Association of America (SCAA), a trade association of industrial environmental remediation contractors, government agencies, and consultants.

Over the years Mr. Katz has prepared approximately a hundred spill prevention control and countermeasure (SPCC) plans for major oil companies, solvent distributors, and other industrial clients. These plans are similar in form and intent to today's environmental audits. Currently he is preparing environmental audits as an independent consultant. He is registered in Illinois as a provider of radon detection services, and he has completed the training courses to be an asbestos inspector and asbestos project designer.

In addition to the above professional activities, Mr. Katz is teaching environmental science for nonscience majors at Roosevelt University and Oakton Community College. After his third year as an adjunct faculty member at Oakton, he was awarded the Ray Hartstein Medal for "Outstanding and Professional Excellence in Teaching."

Acknowledgments

I gratefully acknowledge the patience and understanding of my wife of many years during the writing of this book. I appreciate the careful reading and constructive criticism of my daughter and my son, both of whom improved the readability of my manuscript. And I appreciate the opportunities given me by Oakton Community College in DesPlaines, Illinois, and by Roosevelt University in Chicago to teach this subject. Teaching has been a new career for me, one of challenge and great fun.

Introduction

This book is not intended to be a textbook. It is designed to impart basic information about our environment to people of any age who have little or no scientific background. It is for people who want to know something, but not too much, about the issues that make environmental headlines today, among them famine, the ozone layer, the "greenhouse" effect, and global warming; why a confrontation exists between loggers and conservationists; the importance of the tropical rain forests; the impact of exploding population on available resources; and similar topics. Hopefully some readers will be stimulated to want to know about some of these subjects in greater detail. For them there is a listing of references at the end.

Many of these issues are highly emotional, because they are concerned with the very survival of life on our planet Earth. Both sides in many of the disputes are passionate in their beliefs. On one hand, one's livelihood, the ability of someone to provide for his or her family, is threatened. On the other hand, the existence of our planet is threatened, though at longer range than the immediate need to put food on one's table.

Somehow we must find a way to manage our world that satisfies the needs of all sides of the many disputes that involve environmental issues. We must find a way to provide a "sustainable" earth. It is my hope that this book will provide those not directly involved with enough background to judge the issues for themselves.

In the chapters that follow, many of the ideas are repeated several times, not always in the same form. This repetition is to emphasize the importance of some of the basic "rules" that control life on our planet. Scientists use many specific terms with generally accepted definitions to make communication easier. I have used many of these terms throughout the book. The first time any such word is used it is printed in italics to tell you that is a word you should try to remember. There are a lot of them, so there is a glossary at the end of the book to help you recall the meaning of the word when it reappears later.

Part 1

Matter, Energy, and Population Growth

Chapter 1

The Earth-Sun Relationship

Introduction

We refer to this small planet on which humankind lives as "Mother Earth." If that is true, then we should refer to the star that gives us life itself as "Father Sun." All the energy that sustains life on earth comes from our sun. For millions of years, sunlight has provided the energy that maintains the temperature of earth within a range that life can tolerate. Sunlight provides the energy that plants turn into chemical compounds which provide food for higher life forms. Over millions of years the unused plants have gradually become buried deep below the surface of the earth, and under heat and pressure have evolved into the fossil fuels (coal, oil, and gas)

3

that we extract from the earth for our use today. Sunlight provides the energy that evaporates water from our oceans, creating atmospheric moisture that returns to earth as fresh water when it rains. Sunlight warms our fragile atmosphere, changing the air pressure and creating wind as a result. If the sun "went out," life on earth would cease. If the sun increased its output by even a fraction, life on earth would cease. We live in a delicate balance, totally dependent on the sun for all the energy we need and use.

Law of Conservation of Energy

Energy is indestructible; energy is neither created or destroyed on earth. That fact is known as the *Law of Conservation of Energy* or the *First Law of Thermodynamics*. Energy is created during a nuclear reaction, according to the famous Einstein equation $E = MC^2$, which says that energy and mass (matter) are interchangeable. The energy output of the sun is a result of nuclear reactions that continually convert mass in the sun to energy. We humans are now producing electric power by nuclear reactions on earth. However, compared to the tremendous output of the sun, the amount of nuclear energy produced on earth is so small, and the amount of mass used up in producing it so much smaller, that for all intents and purposes the Law of Conservation of Energy holds true.

Of the total amount of the sun's energy falling on the earth during daylight hours, some is stored in water (mainly the ocean) as heat, some evaporates water into the atmosphere, some warms the land, some is stored in plants in the forms of chemical compounds by the process of photosynthesis, and some warms the atmosphere. When atmospheric moisture condenses into clouds and rain, the heat used to evaporate the water from the oceans (and lakes and rivers) is released into the atmosphere. When organisms use the stored chemicals, mainly "sugars," as food, they also release heat. In humans this results in an average body temperature of 98.6° F.

Heat on the earth, which is at a very low temperature compared to the sun, is distributed unequally. Overall, however, the amount of energy received from the sun must be in balance with that stored on the earth and lost back to space, for the earth's average temperature to remain constant enough to maintain life. There are many local, widespread, short- and long-term variations in the amount of heat retained and radiated. Some of the energy is reflected back into space from clouds, from snow and ice, and from light-colored land surfaces. Various ingredients in the atmosphere, such as methane and carbon dioxide, trap some of this reflected (radiated) heat by what is commonly called the *Greenhouse Effect*. Over a long period of time, when everything is in balance, the earth maintains an average temperature that controls the various climatic zones on its surface. A major scientific question today, which will be discussed in some detail later, is whether the increased amount of carbon dioxide we spew into the air, primarily from burning of fossil fuels, is slowly causing that average temperature to rise. If this is true, the implications of the impact of such a rise are startling.

Entropy

Energy always flows from a highly "organized" or concentrated state to a lower, more "disorganized" or random state. This is known as the Second Law of Thermodynamics. Stated another way, energy always runs downhill. Generally this lower form of energy is heat. For example, the energy in a lump of coal (potential energy) is easily available by burning the coal. The heat produced eventually ends up at low temperature in the environment and is generally unavailable for use. One cannot get back to the higher energy state without using more energy than one obtained going from the higher to the lower level. Take, for example, a car battery. Electricity is produced in a battery by a chemical reaction. The chemical compounds are a stored form of energy. They react to give off electricity, and they form different compounds that will no longer react. When a car battery is "charged" by introducing electricity into it, the chemicals change back into their original form. But it takes more electricity to get back to the charged condition than was available for use. The additional amount shows up as heat; the battery gets warm when it is charged.

Disorganized or random energy, which is less available for use, is known as *entropy*. If the sun were not continually adding energy to the world, we would eventually come to a condition of total entropy (no available energy for use) and the world would be dead. That is why there can be no such thing as perpetual motion. Even with the most perfect bearings man can make, there is always some friction, however small, which generates heat, which in turn is lost for available use. So energy in some form must be added to maintain the "perpetual" motion. There is no "free lunch."

When one considers that only a tiny fraction of the sun's energy reaches the earth, and how much energy the earth uses to keep "running," one can realize how gigantic is the energy output of our sun. And our sun is only a very small star among the uncountable number of stars in the universe.

Law of Conservation of Matter

A fact similar to the conservation of energy is that matter can never be created or destroyed—only changed in form. This is known as the *Law of Conservation of Matter*. When we burn a lump of coal to produce heat, we combine oxygen in the air (a gas) with carbon in the coal (a solid) to produce another gas, carbon dioxide. The weight of coal plus the weight of oxygen equals the weight of carbon dioxide produced. When a fallen tree in a forest is decomposed by microbial action, the tree disappears from view; the debris formed equals the weight of the tree.

Hence there is no "away." We change matter around from one form to another, using up energy or producing heat in the process. It is these changes—which are mostly cyclical in nature—that sustain the living organisms of the world, from the tiniest microbe to man. What we do not or cannot use productively becomes "waste." But we can never get rid of waste, because waste is matter, which cannot be destroyed or created.

What we can do is change the form of that waste. We can turn it into useful products, change it from a dangerous form to a less (or more) dangerous form, bury it, or incinerate it. But we cannot eliminate it from the earth. And every time we change its form, we produce low-level energy from high-level energy. The energy of the sun provides the replacement for the high level energy.

It might seem that when we burn coal or oil we produce energy. But that is only because oil and coal are concentrated forms of energy we can use. Overall, considering the energy to make the coal or oil, to extract it from the earth, to transport it to where we want to use it, to make the machines or equipment that burns it, we have used more energy than we get from it. You cannot get something for nothing! You cannot even break even!

What you should remember

1. All energy on the earth comes from the sun.

2. Energy can neither be created nor destroyed.

3. Energy always runs "downhill."

4. Matter can be neither created nor destroyed.

Chapter 2

The Biosphere

Introduction

Scientists use many specific terms with generally accepted definitions that make communication easier. Many of them are finding their way into the media. We're going to use some of those terms throughout the remainder of this book, so we'd better have a general idea of what we mean by them.

Ecology is the study of the interactions between living things, called *organisms*, and their surroundings, both living and nonliving. In other words, it is the study of how our world works, what makes things live and die, what happens to them, and why.

An *environment* is the total of all the external conditions that affect the life and development of living organisms, including humans. In other words, an environment is the surroundings of living things. Some of the surroundings are other living or dead organisms; some are not capable of life.

We live in what is called the *biosphere*. That word comes from "bio," meaning life, and "sphere," meaning globe. The biosphere contains the organisms, alive and dead, that inhabit the thin layer of air, land, and water on or near the surface of our planet. The biosphere extends from about two hundred or so feet beneath the earth's surface to about twenty thousand or twenty-five thousand feet above sea level.

The air portion of the biosphere is called the *atmosphere*. The water portion is called the *hydrosphere*. The land portion is called the *lithosphere*. Everything within the biosphere reacts with everything else. The total of everything within the biosphere is called the *ecosphere* and the study of ecology is the study of how the ecosphere works.

One interesting experiment has been underway in Arizona in this regard. Called Biosphere 2—the earth is considered to be Biosphere 1—it is a nongovernment experiment in which originally eight scientists, four men and four women, lived in an almost completely self-sustaining enclosure covering about three acres. They were supposed to grow their own food, process their own wastes, generate their own atmosphere, just as would an expedition to another planet, perhaps Mars. The only external input, other than communications, was electric power to keep everything running. (On a Mars expedition power might be supplied by a nuclear power plant.)

Biosphere 2, which is a tourist attraction, contains an ocean, a desert, a tropical rain forest, farmland, living quarters, and hundreds of plants and animals. (Those interested in reading a fascinating account of the first experiment can find a reference in the appendix of this book.) Much controversy arose during this experiment. There were charges that the conduct of the experiment was not self-sustaining, and a number of things did not work out as planned. Nevertheless, the scientists did complete their prescribed time in Biosphere 2. The experiment is still underway, with substantial changes in administration and direction.

There is so much ecology for scientists to study that some kind of structure must be devised to classify the areas to be studied. If you make a list of all kinds of matter, from the very smallest to the largest, from subatomic particles to the entire universe, the study of ecology covers the range from individual organisms at the small end, to the entire ecosphere at the large end. In between are groups of the same kind of organisms called *populations*. Every population exists in a place where conditions favor it. This place is called a *habitat*. A habitat may include several populations, which together form a *community*. A community and all its nonliving physical and chemical surroundings are called an *ecosystem*.

Ecosystems

Ecosystems are of two kinds: *aquatic* or those in the hydrosphere, and *terrestrial* or those on land.

Aquatic ecosystems include areas such as ponds, lakes, the oceans, reefs, estuaries (the mouths of rivers) and wetlands (areas covered with water most of the time). The difference among aquatic ecosystems depends on how deeply the sunlight penetrates, how warm or cold the water is, and how much food is dissolved in the water.

A simpler word for a terrestrial ecosystem (which is kind of a mouthful) is a *biome*. The characteristics of a biome depend entirely on the amount of water (rainfall) it receives and on the temperature.

Deserts

A biome with rainfall less than about ten inches a year, and where more water evaporates than falls as rain, is called a *desert*. Tropical deserts are hot all the time and exist mostly near the equator. Temperate deserts exist in areas that are hot in summer and cool in winter, such as the Sonoran Desert in Arizona. Polar deserts exist in areas which are cold in winter and warm in summer. Parts of Northern Canada and Asia fit this description. (The far Arctic and Antarctic regions, which are snow-covered all year long, are technically deserts.) The common characteristic of all deserts is shortage of rainfall. The plants and animals that survive in deserts have adapted to this situation.

Savannas

A biome with more rainfall than a desert, sufficient to support the growth of grass but not enough to support trees, is called a *savanna*. As with deserts, there are tropical, temperate, and polar savannas. Tropical savannas have high temperature, dry seasons in summer and winter, and a lot of rain in spring and fall; they occur mostly along the equator. Temperate savannas are the main agricultural areas of the world, found mostly in the interior of all continents. Polar savannas occur in the Arctic, where precipitation occurs mainly as snow. These areas are also known as *tundra*.

Forests

Forests occur where there is sufficient water to support the growth of trees. Tropical forests occur where there is a relatively warm climate with little variation in temperature during the year, a lot of rain nearly every day, and very high humidity. Such a rain forest has more kinds of plant and animal life than any other terrestrial ecosystem. Rain forests are easily destroyed, something we will discuss in more detail later on. Temperate forests grow in areas with four distinct seasons per year, combined with moderate temperatures. Trees in such forests are mostly *deciduous*, which means they shed their leaves every year at the end of the growing season. It is possible to have a rain forest in a temperate area, as on the Olympic Peninsula in Washington State. Polar forests are often called *boreal forests,* or sometimes, *taigas*. They are found in subarctic regions, and consist mostly of evergreens.

What you should remember

1. The world is divided into three different interrelated areas: air, land, and water.

2. Scientists have standard words they use in discussing the world and its parts, and the living and nonliving ingredients of the world.

3. The character of land areas and what lives there is determined by the amount of rainfall they receive and the temperature.

Chapter 3

The Earth's Cycles

Introduction

The energy from the sun is changed into usable form by a wonderful and remarkable series of cyclical chemical processes. The primary operators of those processes are green plants of all kinds—from algae to trees. All the cyclical processes are interrelated and interdependent.

The carbon cycle

The most important of the many cycles is the one known as *photosynthesis*, which means "put together from light." Another name for this process is the *carbon cycle*. Plants absorb carbon dioxide and water vapor from the air and energy from the sun, and turn them into chemical compounds known as carbohydrates or sugars, giving off oxygen in the process.

$$(\text{solar energy}) + 6CO_2 + 6H_2O \rightarrow 6O_2 + C_6H_{12}O_6$$

The carbohydrate in this equation is one of many thousands of such compounds produced by the process of photosynthesis.

Plants are thus the *primary producers* in the world. Various organisms—including humans—use those plants for food. They take in the oxygen given off by the plants, break down the sugars into other chemicals that are stored in the consumer's body, give off heat to maintain the organism's living temperature, and emit both water vapor and carbon dioxide back into the atmosphere.

$$C_6H_{12}O_6 + 6O_2 \rightarrow 6H_2O + 6CO_2 + \text{heat}$$

As you can see, photosynthesis is part of a reversible process. The reverse direction is called *respiration,* when cells perform this function in a controlled manner in the body of an organism, or oxidation, when the fuel is burned in a less controlled manner, such as burning a log in a fireplace. In humans the food we eat combines with the oxygen we take in when we breathe; the heat produced maintains our bodies at a "normal" temperature of 98.6° F.

Other cycles

Organisms need more than carbon to exist. Nitrogen in the air is turned into water-soluble compounds by the action of lightning, and by the action of certain kinds of bacteria that exist on the roots of various plants. These soluble nitrogen compounds are taken up by the plants and turned into compounds that are called proteins. These proteins are essential in the diet of animals. When the plants are consumed, the proteins are used to form still other proteins in the consumer. Eventually the consumer either emits nitrogen-containing compounds or dies and is decomposed, thus returning the nitrogen to the soil for reuse. Similar cycles exist for all the other chemicals needed for life. Calcium and phosphorous, for example, have use cycles like those for carbon and nitrogen.

The hydrologic cycle

There is one additional cyclical process that does not produce usable food energy directly, but without which none of the other processes would work. It is called the *hydrologic cycle*, or the water cycle. This is a physical cycle rather than a chemical one.

Most of the water on earth is not usable because it exists in the oceans and is therefore too salty, or it is tied up in the polar ice caps and, therefore, not available in liquid form. The sun's energy causes water to evaporate into the atmosphere, some from freshwater rivers, lakes, and ponds, but mostly from the oceans. The water thus evaporated is fresh, like distilled water; any solids, such as salt in the ocean or other dissolved minerals in lakes or rivers, remain behind. As this vapor reaches the colder upper atmosphere, it condenses into clouds and eventually returns to earth as rain.

Most of the rain falls back into the oceans. Some of the rain falls onto land, where it is directly usable in the carbon cycle. Some seeps down through the ground and fills the empty spaces in porous rock. This *groundwater* exists in slowly moving *aquifers*, which can be tapped by wells to provide usable water. Some runs off the land or falls directly into rivers and lakes. Eventually, because of differences in elevation between the oceans and the land masses surrounding them, most of the water returns to the ocean to begin the cycle over again. However, it may take a very long time for aquifer water to reach the ocean.

Recycling: an invention of nature

In the current environmental motto, the "Three R's," are Replace, Reduce, Recycle. In order to lower the impact of our human societies on our environment, we are urged to:

❏ Reduce our use of everything. Car pool to reduce automotive fuel use, as an example.

❑ Replace high-polluting products with low-polluting products. Change to solar heating instead of burning fossil fuels, for example.

❑ Recycle products instead of just throwing them away. Re-use paper instead of cutting down trees to make new paper, for example.

Recycling, particularly, has caught on as a great new idea. Laws have been passed making recycling mandatory. Ad campaigns tout products as environmentally safe because they are made from recycled materials, or because the container can be recycled. In reality, recycling is as old as the world. Mother Nature has been recycling since the beginning of time. For as you now know, there is no such thing as throwing anything "away," because there is no "away"!

Plants do it

You have read that plants, using energy from the sun, combine carbon dioxide and water to form chemical compounds called carbohydrates, giving off oxygen in the process. They also, depending on the characteristics of the particular plant, form roots, bark, flowers, fruit, seeds, and other essential parts the plant needs to grow and reproduce. Plants also utilize many other elements besides water and carbon dioxide, and they produce many other compounds besides carbohydrates. Scientists call this process *synthesis*.

The weight (scientists call it *mass*) of water plus carbon dioxide, minus the oxygen given off, equals the weight or mass of carbohydrates formed. As the process continues, the plant increases in size, but the increase in mass of the plant is exactly balanced by the mass of water and carbon dioxide taken up, less the mass of oxygen given off.

Eventually something happens to the plant. It may stop growing and die because water or carbon dioxide become unavailable, or because it is cut down. Then the plant starts to decompose. Small animals may gnaw into it, bacteria starts to grow on and in it, and eventually the plant breaks down into many different forms. Some parts become nutrients for other organisms, which then grow and eventually die, decomposing in their turn.

Instead of dying from lack of nutrients, some plants may suffer a sudden end because they are eaten as food by larger organisms. We humans eat many kinds of plants, or parts of them, in our daily food intake. In turn, we eventually die, and decompose.

Trees may be cut down and sawed into boards that are used to build a house. The house may stand for many, many years, but eventually—perhaps several hundred years later—those boards will rot away. Or the house may burn down, turning those boards into ashes and smoke.

Nevertheless, though it may appear that some matter is destroyed, it is only the form that is destroyed. The total amount of matter remains the same. It has only been changed from a solid to a different solid, or a gas, or a liquid.

Changes in matter

There are two kinds of changes, physical and chemical. When ice melts to water, or water freezes into ice, or water is boiled into steam, the three forms of ice (a solid), water (a liquid), and steam (a gas) are only physical changes. When wood burns or iron rusts, while the overall mass remains the same, different compounds are formed. These are chemical changes. Scientists call the study of physical changes "physics." They call the study of chemical changes "chemistry."

Very careful experiments that measure the total mass of all matter involved in either a physical or chemical change reveal a "material balance." In other words, there is never any gain or loss of matter. Everything always balances. What goes in equals what comes out. You already know that scientists call this the Law of Conservation of Matter.

What you should remember

1. Water cycles between the oceans and the land because of evaporation caused by energy from the sun, returning in the form of rain.

2. Plants take up water, carbon dioxide, and energy from the sun to form carbohydrates by the process known as photosynthesis, giving off oxygen in the process. Consumers of those plants take in the oxygen, break down the carbohydrates into usable energy, and give off carbon dioxide and water vapor in a never-ending cycle known as the carbon cycle.

3. Nitrogen needed by plants to make proteins comes mostly from the action of certain bacteria in soil, or the action of lightning, to form soluble nitrogen compounds the plants can use. Consumers return the nitrogen to the soil as a result of decomposition when they die, or in waste products secreted by them.

4. There are other cycles involving calcium, phosphorous, and other elements that operate in similar fashion.

5. All the above cycles are essential to support life on earth.

6. Recycling is a normal part of nature.

Chapter 4

Energy
and
Matter Flow
in
Ecosystems

Introduction

Almost all living organisms obtain the energy they need as part of the carbon cycle. With the exception of a few special organisms, the process of photosynthesis provides the energy required to support an ecosystem. The material produced by this process is plant material of various kinds. Living organisms use this plant material as the starting point that supports a complicated mixture of many kinds of biological species, which taken together is called *biomass*. Commonly biomass is considered to be vegetation, such as grass, trees, or farm crops; it really includes all biological species, including humans, insects, animals, fish, birds, bacteria, and any other living organisms.

The amount of biomass produced in a given ecosystem is called its *productivity*, which is measured by the total amount of everything produced in that ecosystem in a given period of time. Productivity can be reported as the amount of wheat per acre per season on a given farm, or the amount of lumber produced in an entire country per year. Obviously productivity can vary from time to time and place to place.

Energy utilization in ecosystems

The way in which the organisms in an ecosystem share its food supply determines the stability or nonstability of that ecosystem. They ingest plants or other organisms, inhale the oxygen they need to combine with the ingested food, and break that food down into three components: energy, which is used to maintain body temperature; carbon dioxide, which is exhaled; and the various chemicals (starches, fats, protein) that the organisms use to build and repair themselves.

Organisms that feed directly on vegetation are called *primary consumers*. Organisms that feed on primary consumers are called *secondary consumers*. Organisms that feed on secondary consumers are called *tertiary consumers*. As an example, a caterpillar is a primary consumer, the bird that eats the caterpillar is a secondary consumer, and the hawk that eats that bird is a tertiary consumer. There may be many consumer levels in a given ecosystem. These are known as *trophic levels*. "Trophic" comes from the Greek word for food.

Plants are known as *primary producers*. Organisms that use only plants for food are on the first trophic level and are known as *herbivores*. Organisms that use herbivores as food are on the second trophic level and are known as *carnivores* or meat eaters. Some organisms eat both plants and other organisms; they are known as *omnivores*. Some birds, for example, eat both seeds and insects. Some organisms eat other organisms from the same or different trophic levels. Humans, for example, may eat plants, rabbits (first trophic level), and chickens (second trophic level). Chickens are on the second trophic level because they may eat both grain and perhaps also insects. This complex mixture of food supply is called a *food web*.

At each level, energy is lost to the atmosphere in the form of heat. Heat produced in the digestive process beyond that required to maintain the stable body temperature needed for a particular species is radiated (directed to surroundings) or is used to supply the energy to evaporate water (sweat and moisture in exhaled breath). Thus only a part of the high-level energy taken into the organism is used to "run" that organism. Much of it is lost in the form of low-level energy.

Vegetarians: efficient energy users

Moving from one trophic level to the next higher one incurs a loss of about 90 percent of the available energy; only about 10 percent is available for use. Thus, in moving through three levels, 99.9 percent of the energy initially available is lost as heat. That is not very efficient! Also, the size of organisms at the lowest trophic levels tends to be small, so it takes a great many such organisms to provide food for the next higher level. A small bird may eat hundreds of mosquitoes, and it may take tens of those small birds to support one large bird of prey.

As you can see, moving up the food chain is not a very efficient way to use available food energy. Put in everyday terms, this is a strong argument for humans to be vegetarians. It takes a lot of grass and corn to feed the number of cattle required to feed a relatively small number of humans.

Energy does not remain in an ecosystem, but moves through it. It enters by the process of photosynthesis, and ultimately is lost from the ecosystem in the form of heat. This is not a contradiction of the Law of Conservation of Energy, merely an illustration of the Second Law of Thermodynamics—that energy always runs downhill, or that entropy is always increasing.

What happens to mass in an ecosystem is different; it always remains. When plants and all forms of living organisms finally die, they provide food for a variety of organisms, both *scavengers* and *decomposers*. Scavengers, including many insects and some higher organisms, use the dead bodies as food, breaking them down into smaller pieces. Decomposers, mainly fungi and bacteria, further break down the residues into materials that can then be used as nutrients by plants to start the food chain over again. According to the Law of Conservation of Matter, the mass in an ecosystem always remains, though it may be in many different forms.

What you should remember

1. The stability of an ecosystem depends on how the organisms in it share the available food supply.

2. Biomass includes all the biological material produced by an ecosystem.

3. Energy passes through an ecosystem.

4. Matter (mass) remains within an ecosystem.

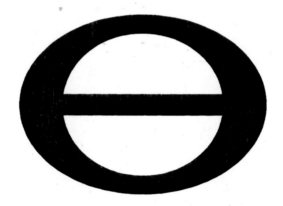

Chapter 5

Population Growth
and the
Ecosystem

Introduction

In the absence of restrictions on growth, all organisms, whatever the species, tend to follow a similar growth pattern. It is only because restrictions exist that some kind of balance can be achieved between the population of a particular species and the resources required to maintain that population. The impact of the growth rate of humans on the resources available on earth, and on the other species that share the earth with humans, is tremendous. An understanding of population growth is essential to an understanding of the problems that face the environment in which we humans live.

Population growth patterns

Organisms all tend to reproduce and multiply in a geometric (exponential) fashion. With sufficient food and proper living conditions, this means that populations increase faster and faster with time. If you make a graph of numbers against time, what you get, with no restraints, is something like Figure 5-1.

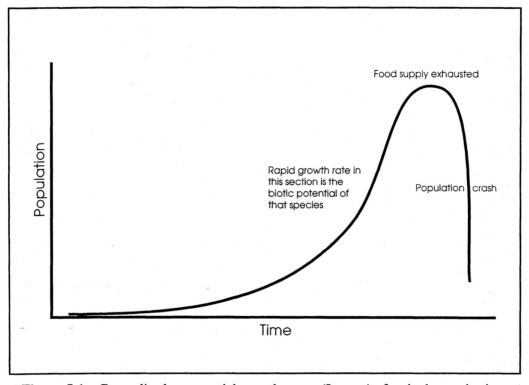

Figure 5-1: Generalized exponential growth curve (J-curve) of a single species in a specific environment.

Scientists describe this as a J-curve, because the shape roughly resembles the letter J. One can demonstrate this growth experimentally by growing bacterial cultures in a dish of food media and counting the number of bacteria under a microscope. In any ecosystem the various species all tend to grow in an exponential manner. But if such growth were to continue forever, everything in the ecosystem would be smothered by sheer numbers of organisms. What usually limits the growth is the amount of available food, though changes in other conditions, such as temperature, moisture, and air, can also serve as limiting factors.

One can estimate the time it takes for a population to double in size by dividing the number 70 by the average percentage growth rate. (This is the same compound interest formula used to calculate the time it takes an investment to double in size if you do not take out the interest earned.) When the available food supply is starting to diminish, the growth rate slows down and levels off. If the food supply is consumed up entirely, the population dies off, or "crashes," and the curve drops off sharply. Sometimes it reaches zero, but usually it stops declining when it reaches a level that the available food can support. Then the growth rate starts to increase again.

Eventually the size of the population will level off at a number that the particular ecosystem can support indefinitely. That number is known as the *carrying capacity* of that ecosystem for that species. Usually a combination of factors—limited food supply, predators, and adverse climate—slows wide fluctuations in population so that the J-curve tends to flatten out into an S-curve as it approaches the carrying capacity.

The reproductive rate along the J portion of the growth curve is known as the *biotic potential* of that species, and it is a species-specific number. The smaller the organism, in general, the higher is its biotic potential. Fruit flies, for example, can multiply so rapidly in a matter of a day or so, that they can be used for genetic studies. Elephants, on the other hand, have a gestation period longer than humans.

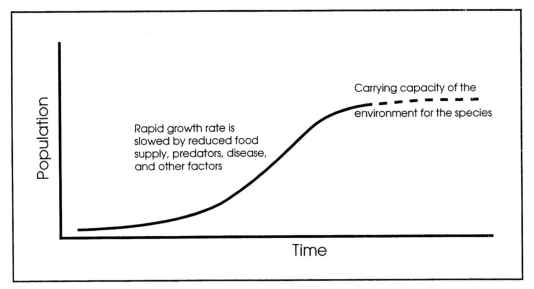

Figure 5-2: Realistic growth curve (S-curve) of a species in a specific environment

Human population growth

Thus, it is with the human population, as shown in Figure 5-3.

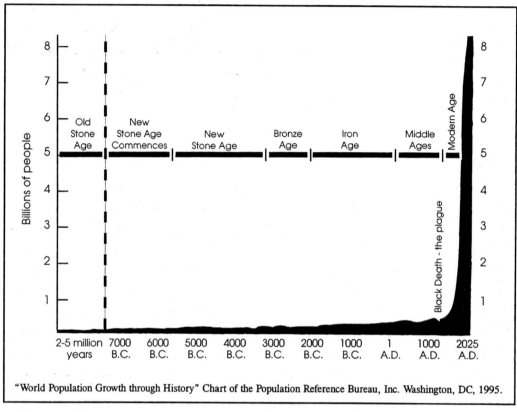

"World Population Growth through History" Chart of the Population Reference Bureau, Inc. Washington, DC, 1995.

Figure 5-3: World population growth

There are several interesting observations to make about all this. First of all, where do the population counts come from? Even in our times, with modern census information, it is apparent that population numbers are only estimates. And, in fact, the numbers differ from one source to another, often by large amounts. Today, who counts the number of people in various tribes in the Amazon Basin, or in the wilds of Borneo, or in strife-torn Africa? And back in the time of the ancient Greeks and Romans, or prior to that, in biblical times, who counted the people in Afghanistan or Tibet? And what about the population counts from 10,000 B.C.?

So some of the numbers, and the shape of the curve going back centuries, must be predicated on the current knowledge of how such curves ought to be shaped. As a result no one really knows at what level the curve starts, but the estimates made today of the rate at which world population is increasing are good enough to cause major concerns about the impact of people on our Mother Earth.

Another interesting observation can be made. Look at the sudden break in the curve between the years 1348 and 1650. That reduction in population was caused by the Black Death, when recurrent plagues killed tens of thousands of people. The world population was low enough and the number of deaths large enough to interrupt the growth curve. Today the death of that number of people would have little observable effect on the growth curve.

The number of humans on earth continues to rise at an alarming rate. It took the world until about 1700 to reach a population of one billion people, about another two hundred years to reach two billion, and only sixty years or so to reach three billion. Today the population is approaching six billion. There are many reasons for this. Advances in medicine and nutrition have steadily reduced the death rate. People in general live longer today than they did a hundred years ago, despite local famines and epidemics. With greater demand for food, agricultural advances have enabled the production of more and more food from the same amount of land, though demand for farmable land increases steadily as well. The impact of wars, which kill off the young and fit of child-producing age, is declining despite our seeming inability to achieve peace, because the proportion of the world population killed in these conflicts becomes smaller and smaller as the population increases.

The extent to which a population exceeds the carrying capacity of its environment is called *overshoot*. The pattern of overshoot followed by a crash is called *Malthusian growth*. In the 1700s, Thomas Malthus, an economist, concluded that this is what would happen to world population. Modern theorists who concur and pattern their objectives for planet management on his conclusions are called *neomalthusians*. Those opposing this idea feel that resources of the world are sufficient to sustain unlimited growth; they are called *cornucopians*, after the mythical "Horn of Plenty," or cornucopia.

Generally, organisms at low trophic levels tend to follow the Malthusian pattern, producing huge numbers of offspring in short periods of time. As you move up the trophic levels, the patterns tend to form S-curves rather than J-curves, slowing as they reach the sustainable level until they come into balance with the carrying capacity of their environment.

The human population has now taken off on the upswing of the J-curve. The impetus to growth has been a combination of factors: discoveries of tools, improved methods of agriculture and industry, and improved ways to utilize available resources. No one really knows what the carrying capacity of the earth is for humans.

What you should remember

1. With no checks on growth, populations of organisms, including humans, increase at an exponential rate.

2. The usual limit to such unrestrained growth is food supply.

3. Every ecosystem has a carrying capacity for the populations in it—that is, the number of those populations that can live there indefinitely in balance with their environment.

4. No one really knows what the carrying capacity of the earth is for humans.

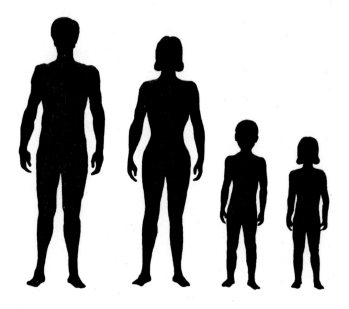

Part 2

The World's Resources

Chapter 6

Agriculture
and the
Food Supply

Introduction

There is sufficient food production in the world today to maintain at least the minimum dietary requirements for everyone. All countries except those below the Sahara in Africa produce enough food for their inhabitants, and some produce a lot more. Despite this, millions of people, mostly children, die because of malnutrition. This is because the available food supply is unequally distributed around the earth. About 2500 calories per day are required on average to maintain good health. Those receiving less than 90 percent of this requirement are undernourished, and those at less than 80 percent are seriously undernourished. Furthermore, a person might consume enough calories in his diet, yet still suffer from *malnutrition* because one may not receive enough of specific requirements such as protein, vitamins, and trace minerals.

Why is there an unequal distribution of food? One reason is that the less-developed countries, in which a small proportion of the people control the land and wealth, need money for industrial development and for arms. So they convert arable (farmable) land from producing food to producing so-called cash crops, such as coffee or bananas (the largest food import into the United States). As a result there is insufficient land for local food production.

About 80 percent of the world food production is consumed by about 20 percent of the population. In the United States farmers are paid subsidies to stop producing some crops in order to maintain prices. Other crops are produced in such large amounts that millions of tons of excess production are kept in storage. Much of the grain produced is used as cattle feed, a very inefficient use of the food value.

Major crops of the world

The three major food crops in the world are rice, wheat, and maize (corn). In high latitudes and in cool regions the staples are potatoes, barley, oats, and rye. In warmer and wetter areas cassava, sweet potatoes, and roots grow well and are staples. Meat and milk are prized worldwide but distribution is inequitable. The less-developed countries raise 60 percent of the world's meat and poultry but consume only 20 percent of it. Fish is another major source of protein, but mankind is already overfishing the oceans.

Where food is grown

Most of the world's crops are grown in the four largest countries, China, India, Russia and the other states of the former USSR, and the United States. These countries hold half the world's population, half the world's croplands, and grow 60 percent of the total food output. The states of the former USSR have more arable land than any other nation. The United States grows the most maize (corn) in the world, 90 percent of which is used as animal feed. (It takes sixteen pounds of grain to produce one pound of edible meat.) India is the third largest producer of staple crops, yet one-third of the country's populace is undernourished. India exports grain, which is used to feed livestock in Europe. China has apparently been able to balance its food production with its controlled population, thus decreasing or eliminating the huge famines that periodically swept that country.

Crop yields

Crop yields have increased faster than world population, without a significant increase in cropland. Most of this is due to the so-called *green revolution*, the development of new crop varieties that, with enough water and fertilizer, produce huge increases in yield. But poor farmers cannot afford the fertilizer, seed, and fuel required to take advantage of the improved seed varieties. And there is much concern among plant specialists that this *monoculture* (growing a single species) might fall prey to a disease, wiping out the crop.

Politics of food distribution

The United States (and other countries) tends to give aid to poor countries, mostly less-developed countries, that support our political policies rather than basing the aid on need. Famines—caused by natural disasters as well as human-caused problems, such as wars—result in mass migration, consumption of feed stock, and many deaths. Poor government planning results in much environmental damage. On the other hand, government planning can overcome the periodic food shortages that result in famine. Such planning has been effective in China, where everyone now apparently has an adequate diet.

At the basis of most food shortages though is overpopulation. Worldwide birth control seems to be a necessity to break the cycle of environmental degradation and inequitable distribution of food supplies.

Soil

Soil for growing crops is one of the world's most valuable resources. It is a mixture of weathered rock, decomposed and decomposing organic matter, and millions of living organisms. Those organisms create the soil's fertility, structure, and tilth (which is the structure suitable for cultivation). It takes thousands of years to build good topsoil, and almost a half of the world's current croplands are being lost faster than the topsoil is being replaced.

Soil loss

The primary cause of soil loss is *erosion*. Other causes are the conversion of arable land to nonagricultural uses, strip mining, poisoning by toxic and hazardous wastes, salinization from improper irrigation, and the misapplication of pesticides.

The world loses about 1 percent of its cropland annually from erosion. Soil erosion exceeds withstandable losses on over 45 percent of the U.S. cropland today. Erosion is caused by wind and water. Either can cause sheet erosion, the peeling off of thin layers of topsoil. Rill erosion occurs when water runs across the land, cutting small channels. These channels can grow to become gullies, which condition is called gully erosion. When exposed, eroded soil dries in the sun, it often breaks down into fine dusty granules. Some of this dust is picked up by winds which carry it well around the world. Here in the U.S., during the mid-1930s, poor farming practice—plowing up the sod, the matted roots of grass, with steel plows—combined with a drought, created the "dust bowl" in Kansas, Nebraska, and Oklahoma.

Irrigation

All plants need water to grow. Overwatering can waterlog soil, killing plants. (African violets and cactus in a home are easily subject to this on a small scale.) Overirrigation can cause a buildup of mineral salts. The most efficient kind of irrigation is *drip irrigation*, developed to a high degree of technical efficiency in Israel. In this method of irrigation, water is delivered just where it is needed at the roots of plants through pipes or flexible tubes perforated with many small holes. Sometimes sensors that measure the plants' water content are used to control when water is turned on or off.

Pests and pesticides

A *pest* is anything living that interferes with the smooth flow of human life. This is probably as good a definition as any, though it might include the teenager next door with a loud boombox. In relation to food, a pest is an insect or a small animal that destroys food supplies intended for human consumption.

It is important to realize that such actions are a normal aspect of the life cycle of those insects or animals, part of their way of reproduction and survival. Insects are important to the life cycle of plants; without them fertilization of crops would be essentially impossible.

A very large amount of food is rendered unusable for human consumption because of pests. The effort to get rid of them has a long history. When a caveman swatted a fly, his hand was acting as a pesticide. Spoilage of food by bacteria, prior to the availability of modern refrigeration, led to the use of salt as a preservative. High concentrations of salt in meat, a pesticide in this use, killed or prevented the growth of spoilage bacteria, the pest.

Today we use many chemical agents to kill pests, and these have worldwide effects. The control of these agents is very difficult. A farmer in Africa using DDT to protect his crops from a locust invasion would not understand an objection based on the fact that DDT dust carried by the wind is settling on the Great Lakes in the U.S. DDT accumulates in fatty tissue as it works its way up the food chain, eventually weakening the shells of eggs laid by birds to such an extent that the shells break when the mother bird sits on them. DDT has been identified as a major factor in lower reproduction rates for many species of birds, and its use as a pesticide has been banned in many areas of the world.

Pesticide dilemma

Because of genetic variations within a species of pest, some individuals within the same species have higher tolerance limits to pesticides than others. Over time the pesticide kills off those individuals with the lower limits, and after a few generations a species resistant to the pesticide has evolved. This may not take much time with insects, because of their very high biotic potential. The result is that higher and higher concentrations of a pesticide must be used, or different and more powerful pesticides must be developed. The potential for harm to humans from exposure to these chemicals, either during application in agriculture or as residues on food, is tremendous.

Medical implications

The failure of antibiotics (pesticide) to destroy certain strains of bacteria (pest) is an example of developed genetic resistance to a pesticide. Many diseases formerly cured rather easily by certain antibiotics are now impervious to the action of those antibiotics. In part this is because of overuse by patients who ask their physicians for an antibiotic, and in part because doctors prescribe antibiotics too readily. Patients stop taking very expensive antibiotics when they begin to feel better after two or three days of a prescribed six- or seven-day regimen, deciding to save the remained "for next time." The result is that some of the bacteria, those with higher tolerance for the drug, are not destroyed. And over time, a drug-resistant bacteria is developed. This is a major medical problem today.

Natural pesticides

Many plants produce natural chemical defenses against pests that may injure them. Planting such crops may deter injurious pests from becoming established. And many predators harmless to humans, such as aphids, lacewings, praying mantis, and others, can be used to control deleterious pests.

Protective laws

The Federal Food, Drug, and Cosmetic Act, as amended in 1958, prohibits the addition of any known cancer-causing agent to processed foods, drugs, or cosmetics. Fresh fruits and vegetables are not covered by this law.

The Federal Insecticide, Fungicide, and Rodenticide Act (FIFRA) regulates the sale and use of pesticides. These laws are difficult to enforce because of the thousands of compounds and processes available, and because many foods today come into the U.S. from abroad. There is poor control of products that are used abroad and inadequate inspection of products reaching the markets in this country.

The Food Quality Protection Act of 1996 combines and clarifies the above two laws, as well as adding coverage of microbial products and "public health" pesticides. Also protected by this act are fresh fruits and vegetables.

Fertilizers

Overfertilization, adding more fertilizer than the plants can take up, can cause pollution from farm runoff, the largest nonpoint source of water pollution in the U.S. That kind of pollution is a major problem in many areas when the potable (drinkable) water supply becomes contaminated. Some methods of fertilizing are less subject to overuse. *Green manure* (plowing under certain crops grown for that purpose), and rotating crops with those that contain nitrogen-fixing bacteria (legumes, such as peas and beans primarily), are two such methods. So is the use of animal manure, though overuse can also cause water runoff problems.

Good farming practices

Less land is now used for agriculture than was used a hundred years ago, because of higher crop yields, and because fewer people farm now than ever before. There is much land that can be brought into cultivation—not in Japan, which is "land poor," but perhaps in Australia, which has the highest per capita amount of available land in the world.

There are many ways to improve cultivation practices, including contour plowing to reduce water runoff, strip farming (mixing crops) to protect soil, and terracing to retain both water and soil. Cover crops, such as alfalfa, rye, and clover, protect the soil and return nutrients. Different ways of tilling the soil can reduce the amount of tilling needed, thus disturbing the surface of the soil less.

Many societies have for generations practiced a kind of self-sustaining farming that conserves and protects their land, using natural fertilizers, fewer or no pesticides, and less energy. Where food production is in balance with population, though yields may be lower this way, costs are also reduced. Soil can remain fertile and productive over long periods of time.

What you should remember

1. There is enough food produced in the world today to provide everyone with a minimum diet, but it is not distributed equally. About 80 percent of the food is consumed by about 20 percent of the people.

2. The main food crops in the world are wheat, corn, and rice.

3. China, the states of the former USSR, India, and the United States grow over half the world's food.

4. Pests are organisms that destroy crops used for human food.

5. Pesticide use is a double-edged sword, saving much food from destruction but exposing humans to potential hazards from toxic chemicals.

6. Overuse of antibiotics has created a serious medical problem because bacteria have become resistant to many antibiotics.

7. Laws to safeguard food supplies in the United States are generally inadequate to regulate imported food.

8. It takes thousands of years to produce good soil for growing crops. About half of this is being lost faster than it is being replaced because of erosion, conversion to nonagricultural use, poisoning from hazardous waste, overirrigation, and poor farming practices.

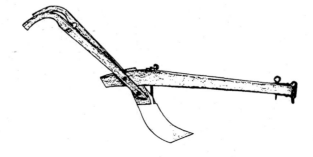

Chapter 7

Land
Managing a Finite Resource

Introduction

The earth's total land area is about 57 million square miles (148 million square kilometers). About two-thirds of the world's land is used by humans. There has been a major change in land use within the last couple of decades, primarily in the conversion of forest to cropland in the less-developed countries south of the equator.

Croplands

Most of the world's food comes from a few species grown on cultivated croplands, which are about 11 percent of the world's lands. Much more could be grown using intensive agriculture methods, but such methods can cause serious environmental problems. The use of large quantities of pesticides and fertilizers creates hazards to many living organisms, including humans. And the growing of single crops, called *monoculture*, exposes that crop to the possibility of being entirely destroyed by disease or an infestation of insects. But croplands are only one of many uses of land in the world.

Forests

Much of the world's land is forest. Forests play a vital role in the world's environment. They provide food and shelter for a large number of the world's species. The trees convert carbon dioxide into carbohydrates (primary foodstuffs for the organisms that inhabit the earth) during photosynthesis and emit the oxygen which is necessary for life. By absorbing carbon dioxide in the atmosphere they help regulate climate. In addition, the tree roots stabilize the ground, preventing erosion and controlling water runoff.

Because of the need to feed the ever-growing human population, nearly two-thirds of the earth's forests have already been converted to cropland, pasture, and settlements. Some have become unproductive wastelands. About two-thirds of the forest that remain are classified as closed canopy; that is, the treetops cover more than 20 percent of the ground underneath. Open canopy forest, or woodland, comprises the remainder, with tree crowns covering less than 20 percent of the ground underneath.

Forest products include industrial timber (lumber, veneer, plywood, and paper) which is slightly less than half of the world's use of wood. The United States is the world's largest importer and second largest exporter of wood. In addition to wood, forests produce oils and waxes, cane, rattan, rubber, sisal, various resins, and essential oils. More than half of the prescription medicines sold in the U.S. contain some natural forest products. Many forest plants contain chemicals that exhibit anticancer activity.

More than half of the population of the world depends on wood for fuel, which means that slightly more than half the wood harvested is used as fuel. About 1.5 billion people have less than they need, and this deficit is expected to grow. The average amount of wood used for cooking and heating worldwide is roughly equal to the amount each person in the U.S. uses as paper products, about one cubic meter (27 cubic feet) a year. Less than one percent of development funds for less-developed countries is allocated to forestry.

Forest management means planning for sustainable harvests of wood. Wood is a *renewable resource*. While most countries plant less than they cut down, many countries have started massive reforestation projects. These include Japan, Russia, Israel, Korea, and China. One problem with replanting is monoculture. Planting fast-growing trees of one species is efficient, but subject to pest and disease. Forest management must also take into account natural causes: volcanoes, fire, insects, and disease.

Tropical forests are the most diverse forests on earth and contain most of the world's diverse species. At the beginning of this century, about 20 million square kilometers were covered with closed canopy forest. Within a ninety-year period about 25 percent has been destroyed and another 25 percent degraded, mostly within the last thirty to forty years.

Rangeland

Rangeland is divided into pasture (enclosed or managed) and open range. Nearly half the world's land is occasionally used as grazing land for domestic animals. If overgrazed, once fertile land can be reduced to wasteland, a process called *desertification*. Most of this is occurring in areas that are arid (contain little water) and overpopulated.

Nature preserves

The idea of setting aside areas for nature preserves is spreading rapidly. Only about 3 percent of the world's land surface is formally protected as parks, nature preserves, and wildlife preserves. The International Union for Conservation of Nature and Natural Resources (IUCN) has identified about 3 billion hectares (7.5 billion acres) worthy of such protection. The loss of forest is a worldwide problem needing a worldwide strategy.

Land ownership

Most of the world's lands are owned by a few people. In Latin America, for example, about 7 percent of the population owns or controls 93 percent of the productive agricultural land. Many governments are trying to break this pattern, but it is a difficult fight. Land ownership in the United States has played a major role in settling this country. Before the European settlers arrived, Native Americans controlled nearly all the land. Now they control less than 4 percent.

Ironically, policies that encouraged land ownership by the settlers to the detriment of the Native Americans probably had as much to do with the development of democracy in the United States as anything else. The Homestead Act allowed settlers to acquire land by living on it and improving it, and thus they gained the right to vote. Some of the acts involving land opened the doors to outrageous fraud. This in turn led to the establishment of many acts designed to protect our natural resources and the establishment of a number of government bureaus and departments that exist to this day. These include, among others, the Bureau of Reclamation, the Bureau of Land Management, the Forest Service, the Fish and Wildlife Service, the National Park Service, and the U.S. Department of Agriculture Soil Conservation Service. Other government agencies have an interest or mandated concern. Among them are the Department of Defense, NASA, the Department of Transportation, the Army Corps of Engineers, and the Department of Housing and Urban Development. A number of acts of Congress, such as the Wilderness Act, Coastal Zone Management Act, even the 1990 Oil Pollution Act, involve protection and management of our land resources.

With so many agencies, conflicts of interest and territorial arguments are inevitable and ongoing. As an example, the Forest Service is at the center of many arguments over logging methods (selective cutting, clear cutting), use of pesticides to manage insect and disease infestations, building of roads and trails, reforestation methods, sale prices for timber rights, and the opening of Alaskan land for commercial use. The Bureau of Land Management is involved in controversy about grazing rights and fees. The National Park Service is trying to reduce the impact of more and more visitors upon wildlife and forest resources within the parks.

What you should remember

1. Forests cover much of the earth's land area and play a vital role in regulating climate and in providing food and shelter for wildlife.

2. About two-thirds of the land is used by humans, for growing crops and as forage and rangeland.

3. Management of land in the United States is regulated by a large number of government agencies, often with conflicting interests.

4. Conversion of forest to cropland has increased dramatically in the last few decades.

5. Only about 3 percent of the world's land is protected as parks or as nature and wildlife preserves.

Chapter 8

Air
Understanding Weather and Climate

Introduction

When the earth formed, the atmosphere was comprised mostly of hydrogen and helium, vastly different than it is now. These are light elements, which over eons of time "evaporated" into space. Insofar as is known, our present atmosphere is unique in our galaxy. It was formed by living organisms that generated the oxygen and carbon dioxide which exist today. The composition of the atmosphere now is about 79 percent nitrogen, 20 percent oxygen and 1 percent other gases, including carbon dioxide, ozone, water vapor, and the rare gases neon, argon, krypton and others.

Atmospheric layers

We live in an ocean of air. Most of it, over 75 percent by weight, is in the *troposphere*, which extends about 10 miles high over the equator to perhaps 5 miles high over the poles. This air, cumulatively, is heavy, exerting a pressure of 14.7 pounds per square inch at sea level. To prove that air has weight, think of what happens when you drink through a straw. Sucking creates a lower pressure in your mouth, and the higher air pressure outside forces the liquid up the straw into your mouth. Above the troposphere is the *stratosphere*, extending up to about 30 miles. It is similar in composition to the troposphere, but it contains a lot less water and a lot more *ozone*, which is produced by lightning and solar radiation.

The ozone layer

The stratosphere is where the much-discussed *ozone layer* exists. Ultraviolet radiation arriving on earth from the sun would cause many human health problems, mainly skin cancers, if nothing diluted its effect. The layer of ozone in the stratosphere reduces the amount by absorbing some of the ultraviolet rays. Plants and animals are also affected by ultraviolet radiation; the ozone layer protects them as well. This subject will be discussed further in Chapter 17, "Air Pollution."

It is known from experiments that the temperature decreases as elevation increases through the troposphere. But then temperature starts to rise as elevation increases through the stratosphere. The stratosphere is relatively calm; therefore, dust and ash from volcanic eruptions (or nuclear explosions) can remain suspended for a long time. If that dust is highly reflective, it has a major effect on solar radiation reaching the earth's surface. After the volcano Tambora in Indonesia erupted in 1815, that year became known as the "year without a summer," because so much sunlight was reflected back into space that the earth was noticeably cooler and darker.

Above the stratosphere is the *mesosphere,* reaching up to about 50 miles, in which the temperature again falls with increased elevation. Above this is the *thermosphere*, which extends up about 1000 miles, and is so rarefied that gases easily ionize under the influence of solar and cosmic radiation. The lower part of this layer is called the *ionosphere*. It is here that the bombardment of charged particles from the sun causes the emission of visible light, creating the auroras known as the northern and southern lights.

Weather

Weather is a description of the physical conditions of the atmosphere, including wind, temperature, pressure, and moisture. *Climate* is a long-term description of weather in a particular area. All weather is driven by energy from the sun and is primarily caused by transfer of energy in the form of heat. The sun evaporates water, mostly from the oceans, and that heated air expands, becomes lighter (per volume), and rises. Winds move the rising air over the earth's surface. When it cools in the upper atmosphere, the air becomes heavier and sinks. The relative humidity increases to saturation, at which point the moisture condenses into rain, giving off the heat of vaporization absorbed in the ocean.

All of the energy falling on earth must be balanced, because of the First Law of Thermodynamics. So the energy released from the condensing moisture is either reradiated back into space (about 70 percent) from the land and sea or directly reflected into space from clouds or land surfaces. The term used to describe reflectivity is *albedo*. Something that has a 50 percent albedo has half of the light falling on it reflected.

Greenhouse effect

The *greenhouse effect* is caused by some of the heat energy that is radiating into space being reflected back onto the earth, which tends to raise the earth's temperature. All kinds of measurements and conjectures and computer models have been made, but no one knows for sure if the rising trend (which is real) is abnormal, or something that has occurred many times over millions of years.

Among other things, weather is caused because solar energy doesn't strike the earth evenly. At the equator it is mostly straight down, while at the poles it strikes at an angle, passing through much more atmosphere as it does so. The earth is also tilted on its axis of rotation, so part of the time the poles get no sunlight at all.

The rotation of the earth creates deflections caused by drag, called the Coriolis effect, and the air flow has several cellular patterns, called Hadley cells, which cause prevailing winds at various latitudes.

Meteorology

The study of this complicated subject of weather and climate is the science of meteorology, which does not have much to do with meteors. Weather has a tremendous impact on health and many human activities, especially agriculture. The prediction of future weather is of tremendous interest; every radio and television station has weather reports, and there are cable television stations devoted entirely to this subject. Boulder, Colorado, is home to a government research facility that is devoted to the study of weather. It is well worth a visit.

Storms

Air movements contain staggering amounts of energy. Wind speeds may reach well over 100 mph, and during cyclones (circular storms over water) the energy fed by heat released from condensing water vapor can reach highly damaging conditions. Over land, without as much water to fuel the storm, a tornado can be much smaller but equally damaging over a smaller area.

Climate prediction

Attempts to study climate over long periods of time have not been very successful, despite more and more sophisticated computer models. Data from long ago, from ice cores and tree rings, extend back only a few hundred or thousand years. Study of geological sediments may go back millions of years, with only general conclusions about climatic changes. The new mathematical study of *chaos*, or the condition of complete disorder, may be able to shed some light on this. One major concern is the impact a nuclear war would have on the earth—primarily because of its effect on weather, an even greater impact than its immediate production of deaths from radiation and blast.

What you should remember

1. As far as is known today, our atmosphere is unique in our galaxy.

2. The atmosphere is about 79 percent nitrogen and 20 percent oxygen, and it exerts a pressure at sea level of 14.7 pounds per square inch.

3. All weather is caused by energy from the sun.

4. The energy falling on earth from the sun must be balanced by energy radiated from the earth back into space, to maintain an average temperature on earth. The greenhouse effect is an increase in the earth's temperature caused by interference with the heat being radiated away from the earth.

Chapter 9

Water

Plentiful, Yet Precious

Introduction

Of all the resources available on earth, water occurs in the greatest quantity. It covers more than 70 percent of the surface of the earth and performs many functions essential to maintenance of life. Most water is salty; only about 3 percent is fresh. The bodies of all animals and plants are comprised of more than 50 percent water; more than 70 percent of the human body is water. Without water, life would cease to exist on earth. Hundreds of daily activities require water to operate.

Fresh water

That small 3 percent of all water that is fresh water isn't spread evenly on earth. Some areas have too much (or more than they need) and others have far too little. Much of it is not easily available; it exists under the surface of the earth as *groundwater*. Groundwater can be tapped by drilling wells into what are known as *aquifers*. An aquifer consists of porous rock, gravel, or sand with the water filling the pores.

There is a common misconception that groundwater exists as an underground river, similar in appearance to a river on the surface. An aquifer may be confined in a manner similar to a river, but the movement of water is slow, often extremely slow. Depending on the tilt of the aquifer, and the distance between the highest and lowest point, flow can be as little as a few feet per year, or an much as several hundred feet in a year. Rarely is the movement any faster than that.

A well drilled into an aquifer may be free flowing. If the upper end of the aquifer, which may be many miles away, is higher than the well at the surface of the ground, water will flow out by itself. This is called an *artesian well*. If there is insufficient height or hydrostatic head to create an artesian well, water must be pumped out of the well.

Aquifers may be very small, a few feet wide, a few feet thick, and a few hundred feet long. Or they may be extremely large, such as the Ogallala aquifer that extends from Nebraska to Texas. Eventually, if water is not removed from an aquifer, it will flow slowly until it reaches the surface in a lake, a river, or even the sea.

Water enters an aquifer, or "recharges it," as the result of rainfall seeping down through the ground. If the amount removed for agriculture, for human consumption, and for industry is less than the recharge rate, an aquifer can exist essentially forever. However, if water is withdrawn faster than it can be renewed, an aquifer will be exhausted. In some instances, depending on the strength of the material comprising the aquifer and the ground above it, the structure may collapse, creating a sinkhole. Many sinkholes have occurred in Florida, for example, where a steadily increasing population and accompanying industry and agriculture has created a demand for fresh water beyond the capacity of aquifers to support that demand. Also in Florida, the withdrawal of too much fresh water has allowed salt water from the ocean to seep back into many aquifers, making them unusable for human consumption.

Environmental problems

There are many environmental problems caused by the ever-increasing need for fresh water and man's attempts to meet that demand by building dams, diverting rivers, constructing reservoirs, and channeling water through pipes and other conduits from where it occurs to where it is needed. This is not a new problem. Two thousand years ago the Romans built aqueducts to supply Rome with water. The problem is just getting worse as the world's population increases and the need for water increases along with it.

At the same time, too much water can also create problems. For example, periodic floods, caused by excessive rainfall, storms, and melting snow, create much suffering because of our inability to predict and control them.

The hydrologic cycle

Most of the water in the world is not available for consumption because it is too salty. Yet the oceans are the controlling factor in the world's weather and the source of the fresh water that is available. The hydrologic cycle, as you read in Chapter 2, is a never-ending process driven by energy from the sun, which constantly replenishes the freshwater supply. The sun evaporates water from the oceans (and also from the surface of lakes and rivers) much as heat supplied to a teakettle turns water into steam. The moisture-laden air cools in the higher portions of the atmosphere, condensing the water vapor into clouds and releasing the heat energy put into the water by the sun. Some of this heat is radiated out into space, in accordance with the Law of Conservation of Energy.

Eventually the clouds become so saturated with water that droplets form, falling to earth as rain or snow. Some of the rain recharges the oceans, lakes, and rivers directly. Much, however, falls onto land. Some of this runs off into lakes and rivers, eventually reaching the ocean. Some soaks into the ground. What is not used by plants, or re-evaporated into the air, may reach and recharge aquifers.

The polar ice caps

In the polar regions of the earth, accumulation of snow over millions of years has formed a cap of ice that is thousands of feet thick. Much of the fresh water on earth is tied up in the two polar ice caps, by far the largest amount in the Antarctic ice cap. If all the ice on both ice caps melted, it is estimated that the level of the oceans in the world would rise more than 360 feet, inundating cities and shorelines around the world. All of the polar ice melting isn't likely to happen. But one of the worries connected with the greenhouse effect is that a slow warming of the earth's atmosphere would melt enough ice to create extensive damage to coastal areas, caused by the rise in the surface level of the oceans.

The polar ice caps are frozen fresh water. Many schemes have been proposed to utilize that source of fresh water to alleviate the need for water in areas of the earth in short supply, such as towing icebergs from Antarctica to the Persian Gulf. At least thus far such schemes have been impractical for many reasons, not the least of which is cost.

Freshwater usage

Most fresh water in the world is used to irrigate crops. In the United States, approximately equal amounts are used for irrigation and in industry. Industry use is primarily for cooling. Most home use of water is for flushing toilets.

What you should remember

1. Water is essential for life on earth to exist.

2. Most of the water is salty and exists in the oceans. Only about 3 percent is fresh water.

3. Only a small amount of the fresh water is readily available for use in agriculture, industry, and human consumption. Most is confined in the polar ice caps or in groundwater. Aquifers are a form of groundwater that can be tapped as a source of fresh water.

4. Fresh water is not uniformly available. Some areas have too much, or at least more than they need. Some have too little.

5. Energy from the sun continuously evaporates water from the surface of the oceans (primarily), lakes, and rivers to form clouds. Rain from this process is fresh water, which recharges aquifers, sustains plant growth, creates lakes and rivers, and eventually returns to the oceans.

6. If water is withdrawn from an aquifer at a rate faster than the recharge rate, the aquifer may be lost as a water source.

7. Worldwide, most water is used for irrigation; in industrial countries about equal amounts are used for cooling and irrigation.

Chapter 10

Living Things
Dealing with Species
Extinction

Introduction

Of the estimated 3 million to 30 million species on earth, about 1.7 million have been identified. We humans get all kinds of benefits from these, and we have no idea of what additional benefits we could have from the species not yet identified. We really do not understand these biological resources very well. Until recently we had little effect on our neighboring organisms, but in very recent times we have been a threat not only to them but to ourselves as well. We threaten to eliminate much of the diversity on the planet because of the effects of a rapidly expanding population.

Living things at a glance

Most of the millions of species are small—insects, microbes, and such. We tend to spend time and effort studying only the larger animals and plants, because the benefits of the smaller species are hard to define. North America and Europe are home to perhaps 15 percent of the world's species. Most live in the tropics, which are incredibly diverse. In terms of *biomass* (weight of living matter), the largest number of multicellular organisms is thought to be krill, which exist in the oceans in vast numbers and serve as food for much of the ocean food chain. Humans probably comprise the next most abundant amount of biomass for a single species.

Wildlife

Plants, animals, and microbes that live independently from humans are considered to be wildlife. Much of our diet comes from that source, including most of the seafood we eat. Much of our diet also comes from forest products that have been cultivated. Many wild edible plants have never been cultivated for human food use.

Biomass loss

There are three kinds of biomass loss of major concern: loss of abundance, species extinction, and ecosystem disruption. Extinction is not new. Studies of fossils suggest that more than 99 percent of all the species that existed are now extinct. As genes are transmitted from one generation to the next, changing conditions allow certain species to prosper and cause others to flounder and die out. Most of this occurs slowly. Occasionally there is an event that causes a mass extinction over a much shorter time, but always that opens the way for new species to develop.

Over eons of time, extinction is a natural process. In undisturbed ecosystems the rate appears to be about one species every ten years or so. In this century alone, however, the estimate is that hundreds of species, including microorganisms, are becoming extinct every year, with the prospect of that number growing into the millions.

Humans destroy species by overhunting and fishing, not always for food. Demand for furs and feathers for warmth and decoration, ivory (elephant and walrus tusks) for souvenirs, rhino horns (used in Asia as an aphrodisiac), cactus for decoration, and animals for pets, all lead to a tremendous illegal traffic. High prices make the risk worthwhile to an inhabitant of a less-developed country, especially when one can earn the equivalent of a year's income from one good poaching trip.

Efforts to control pests and predators by bounties, hunting, poisons, and trapping have decimated many species. Often the rationale for these actions is incorrect, and long-term effects are little considered or understood. How this happens is easy to understand if you are a rancher and see (or think you see) livestock being killed by wild animals and threatening your livelihood.

Ecosystem disruption

Much of the problem is also due to destruction of habitat. The explosion in the white-tailed deer population in the United States (some estimates are that the number of these animals is greater now than when the Pilgrims landed) is because the habitat of predators has been destroyed. Introduction of alien species is another problem, because natural controls developed over time do not exist in the new environment. The rapid growth of such an intruder may lead to the extinction of a previously well-established species.

Careless or uncontrolled use of toxic materials, from pesticides such as DDT to lead shot to lead in gasoline, leads to loss of species. Finally, *genetic assimilation* may gradually cause a species to disappear.

Biomass management

Management of biological resources is a highly controversial subject. On one side are those who say we should leave the world alone, and what will be will be. Their opponents argue that we have no right to cause problems without making an effort to solve them. Here in the United States we have the Endangered Species Act (passed in 1973) which attempts to protect those species in danger of becoming extinct. This law has been a cause of much confrontation between environmentalists and business interests. Laws are passed and often not enforced. Greed and fraud still exist, and there is often no real scientific evidence to support one position or another. We just don't know enough to overcome emotion with reasoned logic and facts.

We have recovery programs for some species, some of which appear to do well; others are more uncertain in their outcome. The peregrine falcon is a good example of a recovery program that seems to work. The condor program in California is a dubious one, because no one knows how well condors raised in captivity will do in the wild. Even apparently well-planned programs of habitat protection sometimes do not turn out well, because of factors overlooked or not understood, as with the Kirkland warbler in Michigan, whose wintering ground was as important to protect as its summer habitat.

Zoos are trying to preserve species by sponsoring breeding programs for those not surviving in the wild. But zoo-raised animals have not learned to survive outside captivity, so release programs have varied success. One interesting program has been the effort by Israel to re-establish the animals described in the Bible as stable wild populations.

We have many wildlife refuges in the United States, the largest of which is in Alaska. These are wonderful to visit, but so many agree with that idea that there are problems with habitat destruction because of too many tourists.

World awareness

Awareness of these problems is growing around the world. Germ plasma banks exist for seeds, plants, and animal sperm, but little is known about the problems of long-term storage (50 to 100 years). Prospects for a positive outcome appear to be good, based on the fact that seeds found in tombs in Egypt have been successfully germinated in modern times.

We are left with many questions and few answers. Hopefully readers of this book will develop an awareness that will help them bring about improvements by their personal actions, voting posture, and lifestyle changes, and in what they teach their children.

What you should remember

1. Extinction of species is a natural phenomenon.

2. Human impact has speeded up the process tremendously.

3. Only a small fraction of the estimated number of species in the world has been classified.

4. Habitat destruction is a major cause of extinction today.

5. Management programs are controversial and varied in success.

Chapter 11

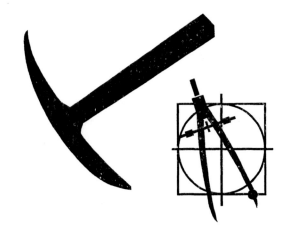

Minerals
A Nonrenewable Resource

Introduction

Minerals are chemical compounds found in the earth's crust. Most of the mineral resources that we use are concentrated in the topmost layer of the earth. They got there by chemical reactions and processes as the earth cooled millions of years ago. They are being renewed, very slowly, by those same natural processes.

The center of the earth is a molten liquid, composed mainly of iron and nickel. Above this is molten rock called the mantle. It is here that the chemical reactions take place forming the mineral resources we need. The thin layer of solidified mantle above all this, the earth's crust (the lithosphere), is where we mine for the minerals we want and need.

The earth's crust

There are three kinds of rocks in the earth's crust: *igneous*, which crystallize out from the molten mantle; *sedimentary,* formed from the settling of deposits in oceans or matter carried by wind, water, and ice, and compacted by pressure; and *metamorphic*, which is both kinds modified by tremendous heat, pressure, and chemical action. Some of these rocks contain metals and some nonmetallic minerals.

Minerals in demand

Most minerals are useful, some are essential to human health and well-being, and some are toxic. Many are both essential and toxic, depending on the dose. Almost everything we use in the modern world depends on one mineral or another. Steel, iron, aluminum, glass, concrete, plaster, gold, silver, and copper are only a few of the minerals we use today; many of these were used in ancient times.

Mineral resources are not replaceable as they are consumed, at least not in the relatively short time frame of human existence; they are exhaustible or *nonrenewable resources. Renewable resources* are those that can be replaced within our time frame, all coming ultimately from the sun's energy and including the biological resources, plant and animal, that we can harvest without exhausting the supply.

Mineral distribution

Mineral resources are not equally available in the world, both in distribution and in ease of recovery. There is a difference between what we can easily recover and use, and the total amount there. Gold is an example. It is sometimes found in a free state, most of which has been recovered. More often it is found in small quantities embedded in rock, which improved technology has increasingly allowed to be recovered in smaller amounts, as the value has risen. Differences in availability of minerals has led to worldwide trade and frequently to political strife between nations.

Cost of minerals

In any market the price of a particular resource depends on the demand for it and the amount available. As demand rises, if there is a low supply, prices rise. That attracts more people to produce the resource, but as prices rise, demand falls off; people find other, less costly materials to do the same job, or they decide to use less. The point at which supply and demand are equal is a state that is rarely achieved and that no one can predict very well.

Most people do not consider the true costs of using a resource. An *internal cost* is the cost to someone who uses the resource. An *external cost* is the real cost in terms of the effect on the world of using that resource. If the market system really operated properly, it would internalize the real costs.

Environmental impact of mineral production

The production of minerals by mining, followed usually by some kind of chemical treatment, has a massive impact on the world's environment. That raises the question of environmental carrying capacity for mineral production. One problem is that 20 percent of the population, more or less, uses 80 percent of the resources. Other than some of the environmentally conscious organizations, no one really tries to include the cost to the environment in calculating what a person or government should pay for that resource.

What you should remember

1. Mineral resources are found in the earth's crust, mostly in the form of chemical compounds.

2. Minerals are nonrenewable in a human time frame. Once used up, they are gone.

3. Minerals are unequally distributed around the earth, resulting in worldwide trade and often political strife to obtain them.

4. The production and use of minerals puts much environmental pressure on the earth, the cost of which is rarely reflected in the market price of those minerals.

Chapter 12

Conventional Energy Resources

Introduction

Energy is the capacity to do work, and *power* is the rate of energy delivery or consumption. Power is always a function of energy per unit of time. Before the development of steam power and the assistance this gave to mining coal, fuel wood was the primary source of the world energy. It still is in many less-developed countries. Today oil is the major source of the world's energy, but the world's proven oil reserves, it is estimated, will be exhausted in less than 100 years. However, there is enough coal for hundreds of years. These energy reserves are crucial because there is a direct correlation between the amount of energy available or used by a country and its standard of living.

The big three

Natural gas is cleaner and cheaper than oil. Reserves of both oil and gas are less evenly distributed than coal. The three (oil, gas, and coal) provide about 90 percent of the world's energy requirements and nearly that high a percentage of the U.S. requirements. Oil is about 40 percent of both totals.

The developed countries consume approximately 90 percent of the gas, 80 percent of the oil, and 65 percent of the world's coal. In the United States the largest share of energy is used by industry, the next largest by transportation (mostly cars and trucks), and the least (still 35 percent) for heating, ventilation, and air conditioning (HVAC), lighting, and heating water in buildings.

We waste a staggering amount of energy. We flare off methane from landfills, we use huge amounts transporting coal and oil to power plants for generating electricity, and we lose about 75 percent of the energy in oil by refining it into liquid fuels.

Coal

There are four types of coal: lignite (low heat value), subbituminous (higher heat value, but smoky), bituminous (most abundant with high heat value), and anthracite (highest heat value, but limited in supply). Most coal was formed several hundred million years ago from fossilized plant material.

Coal mining is dangerous and expensive. Miners are subject to the disease known as black lung disease. Mine tailings, piles of rock left over from coal mining, are very acidic and have degraded their natural surroundings. Many clear streams flowing through Pennsylvania's coal mining region look beautiful and clear, but are sterile of life. Strip mining is cheaper and safer, but also damaging to the environment.

Burning coal puts millions of tons of pollutants into the atmosphere. Sulfur dioxide from burning coal with sulfur in it is a major source of *acid rain*. There is also a tremendous amount of *radioactivity* put into the atmosphere from burning coal. Bedrock all over the world contains varying amounts of uranium, and the natural decay of this long-lived radioactive element causes most coal to be very slightly radioactive.

Oil

Oil is much easier to use than coal because it is a liquid that can be "mined" and transported much more easily than coal. It also can be refined into a myriad of products, mostly also liquid, that can be used to run engines, including cars, trains, and aircraft. Crude oil also is the base material for synthesizing plastics, medicines, and other useful products. The heavy residue from all this can be turned into asphalt, which competes with concrete for road building.

Unlike coal, which is widespread in the world, oil is not. As one result, the United States is dependent on importing oil in tremendous amounts; we do not produce enough oil ourselves to meet the existing demand. Countries like Japan are almost totally dependent on imported oil. This situation has led to widespread fluctuations in world oil prices and extensive political unrest worldwide.

Oil exists occasionally in pools under sufficient pressure to get a gusher, for the same reason that an artesian water well bubbles forth, except that natural gas pressure may increase the flow. Usually oil has to be pumped out of pores in rock and sand where it is trapped, just as in an aquifer. Domestic production of oil in the lower forty-eight states is relatively small. The discovery of the North Slope field in Alaska has provided reserves for another twenty years or so. Our apparent greatest potential now is the Arctic Wildlife Reserve, which is the object of prolonged political fighting between environmentalists and developers.

Natural gas

Natural gas also exists unequally around the world. It can be transferred long distances via pipeline, and also can be cooled to the point where it liquefies. Liquefied natural gas (LNG) shipments are now being made around the world by ocean tanker. There are major safety problems involved.

Coal gasification

Coal gasification is not new. In the early days of this century, coal gas plants existed in many areas of the country. Street and home lighting from coal gas was common in and around large cities. These early plants created areas of hazardous waste from the tars formed in producing the gas, which were simply dumped into pits near the gas-producing facility. Such areas are still creating some problems today. Modern coal gasification plants could produce large amounts of usable fuel from coal in many areas of the world. But the cost of doing so in an environmentally safe manner is higher than the cost of bringing oil to the market at the present time. Experimental plants operated under government funding have been shut down.

What you should remember

1. The more available energy in a country, the higher the standard of living of that country.

2. Much of the world today still depends on fuel wood for its energy source.

3. The main sources of energy for the developed countries are oil and gas. They are not equally distributed around the world.

4. Coal is the most widespread source of fuel in the world.

Chapter 13

Nuclear
Power

Introduction

The use of nuclear power is at the center of a great ongoing debate. Statistically, it is undoubtedly the safest industry in the world from the standpoint of worker accidents, including manufacture of the components for nuclear power plants and the mining and processing of the uranium ores to make nuclear fuel. Considering the pollution injected into the world from fossil fuel plants and accident and health problems in the coal mining industry worldwide, the nuclear power industry comes out a clear winner. Nuclear plants do not emit sulfur dioxide, nitrous oxides, mercury, particulate matter, or radioactivity into the air. The one major problem, as yet not entirely resolved, is finding a way to safely dispose of the radioactive waste.

The difficulty is that we, and the rest of the world, have not developed to any great extent alternatives to the use of fossil fuels. We are running out of oil and gas, at rates subject to the statistics of whoever is presenting the argument. No one seems to disagree very much about the need to develop sustainable natural energy sources. The argument revolves around the use of nuclear power generation to supplant fossil fuel use until such time as sustainable energy sources can supply our needs. And the main thrust of that argument is that there exists a great potential for immediate deaths and genetic damage in the event of a disaster such as the ones that occurred at Three Mile Island and Chernobyl. Much of the argument is emotional, and often it is based on various interpretations of such information as is currently available.

World demand for energy

The demand for energy, mostly as electricity, grows annually, despite periods of slackening, not only in the more-developed countries but also as those that are less developed start to improve their condition. Gross national product and living standard are direct functions of the availability of usable energy. Electricity is a form of energy that is relatively easy to transport from place to place, though it is not easy to store and generating it is usually an inefficient process.

Nuclear power: United States

Illinois has the largest number of commercial nuclear power plants in use in the United States, most of those owned by Commonwealth Edison. In recent years no new plants have been built in this country, and permits to start construction have been withdrawn or not granted because of environmental concerns. This in not true elsewhere in the world. The U.S. had only about 25 percent of the world's total of 375 reactors as of 1986. We led the world into the nuclear power age, and are now leading the world out of it.

Atomic theory simplified

A simplified look at atomic theory, presented here and in the next three sections, may help the reader understand how electricity is produced using nuclear power.

Atoms are composed of a central nucleus which contains most of the weight of an element. This nucleus is made up of *protons*, which have a positive (+) charge. This charge is balanced by a number of *electrons*, which carry a negative (-) charge. The simplest element is hydrogen, with one proton and one electron. It is considered to have an atomic number of 1, for the one proton in its nucleus. The heaviest naturally occurring element is uranium, with 92 protons and 92 electrons—hence an atomic number of 92. By bombarding elements with high-energy particles produced in various kinds of equipment, physicists have succeeded in producing transuranium elements, with higher atomic numbers. These are all short-lived elements and of more theoretical than practical interest to the readers of this book.

Isotopes

There are in the nucleus of some atoms one or more uncharged heavy particles called *neutrons*, which change the mass of the atom to produce what is called an *isotope*. In hydrogen, for example, one extra neutron forms the isotope deuterium, and two extra neutrons form tritium. A certain amount of each possible isotope occurs in nature in a specific ratio in each kind of atom. Water is normally composed of two atoms of hydrogen and one of oxygen, with the formula H_2O, or more properly, ^1H_2O. But a small proportion that occurs naturally is composed of deuterium, with the formula ^2H_2O, or so-called heavy water. The proportion is small, but the total amount existing in the world's oceans is tremendous.

Isotopes can be highly stable, or they can be *radioactive*, which means that they spontaneously emit high-energy radiation while gradually changing to another isotope of the same element, or to another element entirely. This process is called *radioactive decay*; the isotopes that do this are called radionucleides. The rate at which this occurs is specific for each isotope and is measured by what is called its *half-life*, the length of time for half of the isotope to decay. If an isotope has a half-life of ten years, and you start with one pound, in ten years there will be half a pound left, in twenty years a quarter of a pound, in thirty years an eighth of a pound, and so on. This is a geometric progression like the population growth curves, but in a negative direction.

Some radioactive elements and isotopes are highly dangerous to life because of the damaging effect on living cells from the particles and energy that are emitted. Some are extremely useful, especially in medicine for diagnostic work.

Certain isotopes can split apart to form smaller elements, emitting part of the mass of the atom as energy. The relationship between mass and energy was formulated by Albert Einstein in the famous equation $E=MC^2$, which says that energy is equal to mass multiplied by the square of the speed of light. The speed of light is a very large number, which means that the amount of energy that can be produced from a small amount of mass is tremendous.

Carbon dating

Many elements naturally exhibit radioactive decay slowly over time. The ratio of unstable ^{14}C to stable ^{12}C in the atmosphere is essentially constant, and its half-life is known. The carbon cycle puts carbon into plant and animal tissue at the atmospheric ratio, after which the ^{14}C slowly decays with time into the stable ^{12}C form. By measuring the ratio in plant or animal residues in archeological finds, it is possible to calculate the age of those residues. This process is known as carbon dating.

Chain reactions

Some isotopes emit neutrons when they decay or split apart, and some of those isotopes are marginally stable. When impacted by a neutron, they not only split apart and produce energy in the form of heat but also produce additional neutrons. These neutrons in turn can impact other atoms and cause them to split, releasing more energy and more neutrons, thereby creating what is called a *chain reaction*. If there is insufficient isotope present, a chain reaction will not proceed or will die out. If there is just enough, the reaction will maintain itself. A place has been found in Africa where such a naturally occurring chain reaction apparently proceeded for thousands of years.

Atom bombs

An atom bomb is an uncontrolled chain reaction. If there is a large amount of isotope present, the reaction can proceed explosively, forming an atom bomb. The power is awesome from a small amount of mass. The trick is to keep separate portions of the isotope far enough apart so they do not react, then bring them together fast enough so the heat and energy generated does not push them too far apart and stop the reaction. This is usually done by exploding them together.

The first atom bomb was made from uranium isotope ^{235}U which exists is very small amounts mixed with the more common ^{238}U. Later, hydrogen in the form of tritium, ^{3}H, was used to make a hydrogen bomb. These reactions are called *fission* reactions, because something comes apart.

Nuclear power plants

A nuclear power plant uses the same reaction, but the number of neutrons is carefully controlled by interposing a material that can capture and absorb neutrons, so the rate of neutron production is just sufficient to keep the reaction going. The first such chain reaction occurred under Stagg Field at the University of Chicago.

Fuel for nuclear power plants is made by concentrating the amount of ^{235}U sufficiently so that fuel pellets can be made from the powder. There are several processes for doing this, most developed during World War II and since greatly improved. The pellets are designed so that a controlled chain reaction can occur, but an explosive reaction cannot.

In a nuclear power plant heat from the nuclear reaction is used to generate steam, as is heat from burning coal or oil in a conventional power plant. The actual electricity generating equipment is the same in both kinds of plants.

Radiation's effect on living tissue

Radiation affects living tissue. The amount absorbed is measured by a unit called a rad, a specific amount of radiation per unit of body weight. This is modified by a quality factor Q, determined by equipment and/or activities that may modify the effect of the radiation. The amount absorbed by humans is measured by a unit called a rem, which is equal to rad times Q. The number is very large so meters and measurement badges are calibrated in millirem, or one-thousandths of a rem.

Naturally occurring radiation

There is much naturally occurring radiation from isotopes in the rocks and ground, from cosmic rays (the amount depending on altitude, with less shielding at higher elevations), and especially from *radon,* a decay product of uranium and thorium in the earth.

Fusion reactions

Fission reactions produce large amounts of radioactive decay products, creating tremendous engineering design problems. Nuclear reactions where isotopes combine to form higher elements also give off energy, because the sum of the lower-weight element is a little less than the resulting higher-weight element, the difference appearing as heat energy. This is the reaction that is taking place in our sun, where hydrogen is combining to form helium. This process is called fusion. Despite many attempts to produce a controllable, workable fusion reaction, it has not yet been accomplished. If it ever is, this may be the nuclear power source of the future, because the dangerous by-products are minimal compared to those in a fission reaction.

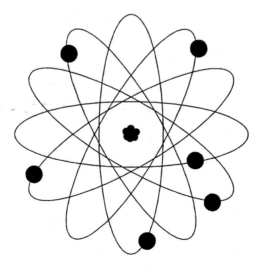

What you should remember

1. The nuclear power industry is far less environmentally polluting than conventional methods of generating power.

2. The worldwide demand for energy, particularly electricity, is growing, especially in the less-developed countries.

3. The United States produces more nuclear power than any other country, but construction of new plants has stopped here.

4. Every chemical element may have one or more isotopes, many of which decay radioactively at a fixed rate or half-life. The ratio of isotopes in carbon atoms, together with their known half-life, can be used in carbon dating.

5. Isotopes can be induced to split into lighter elements by hitting them with neutrons. The mass of the elements produced is slightly less than the mass of the original isotope. The mass lost is turned into energy in the form of heat.

6. A small amount of mass can produce a large amount of heat energy.

7. If the decay products include neutrons, which in turn can cause more atoms to split, a chain reaction is possible.

8. An atom bomb is an uncontrolled chain reaction. A nuclear power plant is a controlled chain reaction.

9. There is much naturally occurring radiation to which we are all exposed.

Chapter 14

Renewable Energy Resources

Introduction

Every year the United States wastes more energy than the rest of the world uses! The problem in using renewable and sustainable sources of energy is purely and simply one of cost, of economics. We can get energy cheaper from coal, oil, and nuclear power, so we do that with little regard for the future our children and grandchildren and great-grandchildren may have to face.

The United States obtains about 95 percent of its power from nonrenewable sources. We could save tremendous amounts of energy by conservation methods, some of which came into play during the oil crisis in the 1970s. Several European countries with higher standards of living than ours use less than half the energy we use. Cheap oil is a political tool in this country, and neither the Republicans nor the Democrats have the courage or the wisdom or the willpower to do much about it.

Solar Heating

There are two kinds of solar heating: *active* and *passive*. Passive solar heating involves building a home to take advantage of the sun, something as simple as installing a greenhouse wall to substitute for a furnace. Roof overhangs shade the windows from summer sun, and south-facing windows allow heat from the sun to warm the house in winter.

Active solar heating uses the sun to provide the energy needed to boil water to form steam to drive an electric generator. There are some projects that utilize mirrors to reflect solar heat for this process. One or two are in this country, one is at the Weitzmann Institute in Israel, and another is in France.

Photovoltaic cells

Photovoltaic cells can convert light directly into direct-current electricity. On a small scale, these cells can operate calculators and other small appliances, using light from household lighting as their energy source. On a larger scale, arrays of cells connected in series use sunlight to produce large amounts of direct current power.

While the high cost of such cells has been an impediment in the past, the efficiency of photocells is improving constantly and costs have been decreasing steadily. Panels of cells can be installed in remote areas where electricity is otherwise unavailable. Swimming pools have been heated and their circulating pumps operated from solar panels. The potential uses are almost endless.

Energy storage

Storage of energy is a problem. Batteries are the usual storage medium. Because an electric motor run "backwards" becomes an electric generator, one storage method uses a motor to pump water to a higher elevation during periods of low demand and lets it run back downhill turning the motor into a generator when additional power is needed. Consumers Power has such a facility at Ludington, Michigan.

Biomass

Probably the largest potential source of energy is biomass. Burning waste *biomass* (tree chips, agricultural residues, waste lumber, and similar materials) produces carbon dioxide, as does burning coal. The difference is that the photosynthesis process that produces those materials uses the carbon dioxide produced, so the net impact on the world's carbon dioxide supply is zero.

Anaerobic (absence of air) digestion by bacteria produces methane from plant and animal matter. Newer landfills are fitted with collection pipes to catch the methane and burn it for fuel to drive an engine connected to a generator to produce electricity. Animal manure is collected and decomposed in anaerobic digesters to produce methane in many small villages in the world.

The largest use of wood in the world is still for fuel. Controlled production of fast-growing trees and selective harvesting is a needed process in less-developed countries, requiring much education. Municipal garbage is burned for fuel in some areas. Agricultural waste is burned as fuel to generate steam in many industries. Bagasse from sugar cane is one example. Peat (sort of an intermediate step in production of coal—incompletely converted plants) has been burned as fuel for centuries.

Some countries are experimenting with using alcohol for automotive fuel, Brazil primarily. The United States has long produced industrial alcohol from corn and beet sugars, and in many states this is blended with gasoline to form *gasohol*. Less than 10 percent ethyl alcohol in gasoline burns well in present engines.

Geothermal energy

Geothermal energy, using the heat in the earth either by pumping water down into heated rock layers or using the natural production of steam and hot water, is being utilized in California and in New Zealand on a large scale. There are serious technical problems with corrosion that must be overcome in installations that employ this process.

Cogeneration

One growing trend in the United States is *cogeneration*—utilizing some of the heat produced by burning coal, gas, or oil, which is presently thrown away unused—to produce both electricity and steam in the same facility. Many industrial plants are now starting to use this process as a way of both reducing costs and conserving energy..

Wind power

The wind has long been used to produce energy—from sailing vessels to windmills that pump water. Several companies in the United States now produce small windmill units to generate electricity for individual homes. Out West, especially in California, or any place where winds are relatively constant, wind farms have sprung up. The cost of producing energy from wind is decreasing and is approaching the cost of energy produced from more conventional power sources.

Hydroelectric power

The potential energy in falling water has long been used to run grain mills and to generate electricity. Niagara Falls is a good example of large power-plant construction on a river. In many areas of this country and the world there are small streams with individual power plants serving small communities. Some people are purchasing abandoned plants of this kind to generate what they need and selling excess production to the larger utilities. Where dams do not have to be built, creating flooding and silting problems, this is a potentially good source of power on a small scale.

Other alternative sources

Other techniques are mostly still experimental, such as Offshore Thermal Energy Conversion (OTEC). OTEC uses the heat in ocean surface water to evaporate a low-boiling liquid which, like steam, turns an electric generator. Cold water well below the surface condenses the vapors, which are recycled in the process. The potential power in tides is huge. In areas where the difference in height between low and high tide is very great, such as the Bay of Fundy, gated dams can allow the water to be trapped at high tide and released through generators at low tide to produce electricity.

Cost

The main impediment to constructing sufficient clean energy-production facilities to supplant present-day fossil and nuclear fuel plants is cost. It will take staggering amounts of money and years of time to accomplish this.

What you should remember

1. There are many sources of clean, nonpolluting energy available, but none in sufficient amounts to supplant present fossil and nuclear fuel power plants.

2. The cost of some clean energy sources is approaching the cost of conventional power generation.

3. Biomass as a source of fuel for power plants is overall a nonpolluting source of energy.

Part 3

Pollution

Chapter 15

Environmental Health and Risk

Introduction

Health is a state of well-being: mental, physical, and social. The greatest health threat for most people in the world is pathogenic organisms; gastrointestinal infections cause more deaths than any other group of diseases, mostly among children in the less-developed countries. Simple rehydration therapy and good food would prevent many of these deaths. The next greatest threat is toxic or *hazardous materials*, which include *carcinogens* (cancer-causing substances), teratogens (abnormality-causing substances, such as thalidomide), and mutagens (substances that affect the genes).

In the United States, control of such materials is under the aegis of the Environmental Protection Agency (EPA).

TSCA

Congress passed the Toxic Substances Control Act (TSCA) in 1976. Under the provisions of this legislation, the EPA has listed several hundred such hazardous materials with permissible discharge limits. Also listed for each material is the LD_{50} rate, which is the dosage that can cause half of test organisms exposed to that material to die within a stipulated period of time, usually 48 hours. Any material not on the list that is flammable, explosive, toxic, or corrosive, is also considered to be hazardous and is reportable if discharged.

Effect of concentration

Many materials that are dangerous at one (usually high) concentration are beneficial at a lower one. This is particularly true of medicinal drugs. Ordinary salt (sodium chloride), which is essential at low concentrations for the operation of bodily processes, can cause high blood pressure at high concentrations in the blood. Yet even high concentrations can have benefits. Bacteria that cause spoilage of some foods cannot live in high salt concentrations, a fact that was and is used all over the world to preserve food where refrigeration is not available.

Risk

It is impossible for humans to avoid contact with everything that can cause problems. In part that is because we all have different tolerance levels for substances to which we are exposed. Even common everyday materials, such as salt, MSG (a naturally occurring flavor enhancer), or any of the large variety of plastics, can cause allergic reactions, sometimes fatal, in some people. Many of these substances are man-made; some are naturally occurring.

Consequently everyone is subject to risk, the chance of something detrimental occurring. An entire engineering specialty known as risk analysis has developed, primarily in the insurance industry but also in manufacturing. Practitioners attempt to quantify risks by the probability of occurrence and the impact if encountered, and to determine the factors that can be modified to reduce the risk. This is known as risk management.

Cost of risk management

Cost is always a major concern. It is a generally accepted fact that the first 90 percent reduction of a risk costs about the first 90 percent of the money, say a million dollars. Then the next 9 percent reduction will cost about the same amount, or another million dollars. Likewise the next 0.9 percent reduction will cost about another million dollars. And so on. At some point a judgment must be made, a trade-off considered, as to whether or not it is better to continue trying to reduce one specific risk, or to spend the money someplace else where one gets more "bang for the buck."

Can there ever be zero risk? Probably not. Often when the cause of an accident is investigated, it is found that there were, say, five elements that had to be present at the same time to cause the accident. Frequently it happens that after arranging matters so it is impossible for all five to occur, there is another accident and it is discovered that there was a sixth factor never considered at all.

Acceptable risk

The decision of where to concentrate efforts to reduce a hazard raises the question of what constitutes an acceptable risk. This subject affects all of us all the time, and what we are willing to accept depends to a great extent on our perception of the risk, how much we know about it, and how much we think we can control it. This applies to many things in our lives: the cars we drive, the foods we eat, the company we keep, the lifestyle we practice, and even the public officials we elect.

As an example of lifestyle choices, riding horseback provides exercise and pleasure, at the risk of being thrown and hurt. Many are perfectly willing to accept this risk, if they can afford to do so. There is probably far less chance of accident and greater benefit to walking briskly three or four miles daily, but many people will not make that choice. Instead they choose to drive for recreation, which costs a lot more (probably) than owning a horse and certainly exposes them to far greater risk of accident.

Diet is another risk, because of the well-proven correlation between many diseases and the foods we eat. We all have choices, but few of us in this overendowed country eat the way we should to control body weight and the risk of various kinds of cancer.

Ethical questions

There are many ethical questions involved in studying the environment, and that study spreads beyond just science into the areas of law, government, history, sociology, and even philosophy. Our government attempts, through the Food and Drug Administration, the Environmental Protection Agency, and various congressional legislative activities, to minimize our risks. Many of these activities require testing, often involving animals. Is this morally and ethically proper? Should we be conducting tests on humans as was done in the concentration camps in Germany and apparently on U.S. soldiers exposed to radiation from the atom bomb tests in this country? Should the data from those experiments, if pertinent, be used to answer questions today?

What you should remember

1. The greatest health threat in the world is pathogenic disease organisms.

2. The next most serious threat is toxic/hazardous material.

3. Everything that we do involves risk of some kind.

4. Risk analysis has developed as a scientific and engineering method of quantifying risks and reducing the probability of accidents.

5. There is probably no such thing as zero risk.

6. The risks we are willing to accept in our lives depend to a large extent on the perception we have of how much control we have over those risks.

Chapter 16

Water
Pollution

Introduction

A pollutant is anything physical, chemical, or biological that adversely affects the quality of water, air, or land. The federal government has established standards for water quality, initially under the Federal Water Pollution Control Act (FWPCA), then the Clean Water Act (CWA), administered by what is now the Environmental Protection Agency. Most states also have comparable agencies, some with standards much stricter than the federal ones. There is a complex, interwoven network of regulations that industry and local governments must meet, all designed to protect you and me.

Point sources

Such standards distinguish between *point sources*, emissions from a well-defined source, such as a sewer outfall, and *nonpoint sources*, such as storm-water runoff. Point sources are controlled by the National Pollution Discharge Elimination System, commonly known as the NPDES program. This program requires facilities to monitor and report their discharges of materials on the EPA's hazardous materials list. These permits require control of infectious agents, biological and chemical oxygen demand, and specific chemical elements or compounds that may be toxic or otherwise dangerous to human health.

Nonpoint sources

One of the major water pollution problems is control of nonpoint sources, primarily from agriculture and storm water runoff. Automobiles deposit millions of tons of residue on the land (road film), much of which used to contain lead from combustion of tetraethyl lead, an antiknock compound. Much dirt is carried along with such runoff, so that sediment and suspended solids comprise the largest amount of water pollution not only in the United States but in most parts of the world. Heat is also a nonpoint pollutant that can be a major problem, primarily from power plants and industrial processes that require the use of cooling water.

Groundwater

Groundwater pollution is a significant problem because it can contaminate a major source of potable water, sometimes with fatal results. Such an incident occurred in Illinois when cyanide plating waste dumped on the ground a few years ago caused illness and death in a farm family not far from Chicago.

Underground storage tanks

There are an estimated twenty million underground storage tanks (USTs) in the United States, primarily for automotive fuel in service stations, but also for home heating oil, farm equipment, and various chemicals. It has also been estimated that 80 percent or more leak. Management of such tanks has been turned over to most states by the EPA. Registration of USTs, usually with a state fire marshal or equivalent, and listing of leaking underground storage tanks (LUSTs) is mandatory. Remedial measures include tank repair, tank removal and removal of contaminated soil, installation of corrosion protection equipment, and installation of monitoring wells around USTs, all designed to prevent migration of leaking material into groundwater.

Oil Spills

Every year in the United States there are several thousand oil spills reported to the U.S. Coast Guard (for spills on navigable waterways), the EPA, or to the various state equivalents of the EPA. Millions of gallons of various kinds of oil are transported by truck or barge, in drums, through pipelines, and by other means. Spills are an inevitable part of an economy based on oil energy. Most of these are small, properly and quickly cleaned up by trained employees of private companies engaged in that business, and they never get public notice.

What does come to public notice are the very large spills, most often of crude oil, from an accident involving a supertanker moving crude from its point of origin to ports in one country or another. The *Amoco Cadiz* accident off the cost of Normandy, and the *Exxon Valdez* disaster in Alaska, are examples of this.

Oil spill legislation

The focus of the EPA has been on prevention, by requiring facilities that could possibly cause an accidental spill into any navigable water of the United States to prepare a Spill Prevention Control and Countermeasure (SPCC) plan. These plans must be reviewed and updated if necessary at regular intervals. Large ocean spills get immediate media attention and create an enormous public outcry at the sight of dead wildlife coated with oil.

As a result of the *Exxon Valdez* accident, new legislation, the Oil Pollution Act of 1990 (OPA90), has established reserves of equipment and maintenance of trained personnel at various points around the ocean shores of this country to respond quickly to spill incidents. Those facilities have not had much use because spills like the *Exxon Valdez* do not occur very often.

Spill management

Because there is no "away," cleaning up an oil spill involves trade-offs. Readers familiar with the Dr. Seuss children's book *The Cat in the Hat* may remember the incident of the spill of pink ink. An approximate conclusion of that story is that to make something clean something else must get dirty, and if you are careless or inefficient nothing gets clean and everything gets dirty.

Dispersing spilled crude into the water column may remove contamination of shorelines and the organisms that live there, but it may put water-dwelling organisms at greater risk than if the oil remained floating on the surface. Mopping up oil with sorbent pads may leave a cleaned surface but produces large quantities of dirty pads that must be buried or burned or otherwise disposed of.

Over twenty years ago a consensus standards-writing organization, the American Society for Testing and Materials (ASTM), established Committee F-20 on Oil Spills (now Hazardous Material and Oil Spills). This brought together representatives of government, industry, academia, and miscellaneous interested parties. Over the years F-20 has produced many standards covering equipment and procedures for controlling and cleaning up oil spills, especially large ocean spills. An industry trade organization, the Spill Control Association of America (SCAA), was established about the same time, to provide a forum for contractors with common interests in oil spill cleanup. Similar efforts and organizations have been established around the world.

Spill cleanup effectiveness

Although there has been some improvement in equipment and methods of trying to manage oil spills, there has been no real improvement in efficiency of cleanup in over twenty years, despite millions of dollars and man-hours spent on the problem. It is, of course, necessary to try to protect property and restore beaches and shorelines in highly developed areas. But in remote areas cleanup efforts frequently do more damage than the spill itself. Disagreements among both the state and federal agencies that are impacted by a large oil spill do not improve cleanup efficiency.

Microbial action will eventually decompose spilled oil, slowly in cold areas like Alaska and much more rapidly in warm areas like the spills in the Persian

Gulf. Many natural oil seeps, such as those in the Santa Barbara, California, channel have been known for years. They are left alone and nature takes care of the cleanup.

In the author's opinion, not shared by many agencies or individuals whose funding and businesses are "improved" by large oil spills, it would be better to let nature handle the spills in remote areas, and spend the huge amounts of money required for cleanup activity on prevention instead.

Sewage treatment

Sewage treatment, widely practiced in the more-developed countries, is not so widely practiced in the less-developed countries. Where populations are small, people simply go into the woods or forest to relieve themselves, and natural bacterial decomposition takes care of the problem. Where populations are large, and sewage is simply discharged into rivers or lakes without treatment, major health problems arise from pathogenic organisms. In 1991, an outbreak of cholera, a gastrointestinal disease, arose in South America. It is still a major problem in Peru and other South American countries, and there have been minor outbreaks and scares in the United States as well.

Septic tanks

In developed countries septic tanks are in widespread use in rural areas and even in urban areas that have not been connected to a central sewage treatment facility. A septic tank settles solid waste as sludge, which must be pumped out at intervals, segregates greases and oils at the top, which similarly must be pumped out, and spreads the remaining liquid over a wide area where it can oxygenate, which kills the pathogens. You can almost always tell where a septic field is working properly because the grass above is always greener than elsewhere. But, as with most generalizations, this does not always hold true. And of course, the pumped-out material must be properly disposed of, usually in a licensed landfill.

Modern treatment plants

Modern sewage treatment plants are highly efficient and often surprisingly odor free. Large solids are removed by screens, for landfill disposal. Small suspended solids are settled out as grit and sludge, which (before there was so much toxic heavy metal included) was and is still often sold as fertilizer. The liquid portion is "digested" microbially to produce a material safe for discharge into watercourses, sometimes after chlorine treatment for final removal of pathogens.

Natural sewage treatment

A new technique being tried experimentally in the United States, using large marshy areas containing certain kinds of reeds and other plants, shows great promise of safely and inexpensively turning sewage into potable water—a technique that could well be adapted in many less-developed countries. The idea is not really new, but widespread use as a replacement for more conventional sewage treatment is relatively new.

Population impact

As population grows in urban areas, sewer systems get overloaded. In many areas sanitary and storm sewers are one and the same, with the result that under storm conditions raw sewage gets discharged into lakes and rivers. When this happens, the high levels of pollution force the closing of swimming beaches, a not unfamiliar occurrence along the shorelines of the Great Lakes.

Biodegradability

The term "biodegradable" has become part of our vocabulary, but what does it mean? The generally accepted meaning is that something biodegradable will break down into harmless decomposition products. However, without the two specifics of time and amount, it is an inexact term. Eventually everything is biodegradable, though the time involved may be very long indeed.

As it is used today in describing consumer products, the term is a buzz word with little useful meaning. With the specifics of time and amount added, it would be possible to make a judgment about the relative biodegradability of various products. Hence "50 percent degraded in 14 days" is an accurate measure of biodegradability, better than "40 percent in 60 days."

What you should remember

1. Water pollution arises from point and nonpoint sources.

2. Point sources are regulated by the NPDES program under the Clean Water Act.

3. Nonpoint sources, from agricultural and storm-water runoff, are not easily controlled.

4. The term "biodegradable" has little meaning unless a time and amount are attached to the term.

Chapter 17

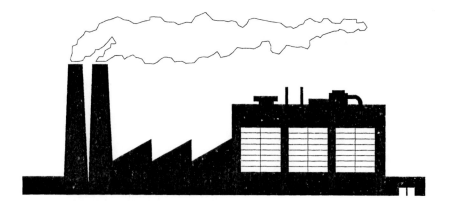

Air
Pollution

Introduction

Air pollution covers the world, carried everywhere by the wind. Some pollutants are natural, from fires, volcanoes, dust, and odors. Most are from human activities. The quantities of pollutants are staggeringly large, amounting to millions of tons worldwide. The capacity of our atmosphere to absorb such amounts without affecting everyone and everything that "breathes" is not limitless.

The composition of pollution from natural and human sources is often the same. Many human-made materials are different, however, and can cause great damage in very small quantities.

Categories of air pollutants

Primary pollutants are those entering the air directly in harmful form. *Secondary pollutants* are those created by chemical or physical changes after they enter the air. Many pollutants are essentially "point source," such as those emitted from a smokestack or automobile tailpipe. They are at least potentially subject to control. Many, called *fugitive*, are nonpoint source and are thus difficult to locate accurately and even more difficult to control.

Critical pollutants

The Clean Air Act of 1970 specified seven *critical pollutants* for which maximum permissible levels in ambient (surrounding) air are established. These are sulfur dioxide, carbon monoxide, nitrogen oxides, particulate matter, hydrocarbons, *photochemical oxidants*, and lead. They constitute the largest volume of atmospheric pollutants, and are the most dangerous to human health.

Sulfur compounds come from natural sources, such as sea spray, volcanoes, sulfur-containing dust, and emission from bacterial action. Only about 25 percent of all the sulfur is from human-related activities, mostly burning sulfur-containing coal. Sulfur dioxide directly affects plants and animals and can be further oxidized in air to form sulfur trioxide, which with water forms sulfuric acid, the main constituent of "acid rain." Next to smoking, sulfur air pollution is probably the most prevalent cause of health damage.

Nitrogen compounds form from burning fuels at temperatures above 1200° F. Nitrogen dioxide is brown in color, giving photochemical smog its characteristic color. Combined with water this forms nitric acid, the other main constituent of acid rain. Nearly all of this in the United States comes from burning fuel for transportation and electric power generation.

Carbon oxides are mostly in the form of carbon dioxide, the bulk of which comes from plant and animal respiration. This is just about balanced by uptake in green plants in photosynthesis. Levels are slowly climbing because of burning of fossil fuels and biomass, amounting to about 20 percent of the total. Carbon monoxide is released from incomplete combustion of fuel. It is colorless, odorless, and dangerous, because it binds to hemoglobin and prevents uptake of oxygen. About half comes from human activities, and two-thirds of that from internal combustion engines.

Metals and halogenated compounds come from mining dust, trace metals in fuels, and waste disposal. The largest quantity is lead, about two-thirds of all metal pollutants, and that comes from burning tetraethyl lead, a compound used to improve the ignition properties of gasoline. Burning leaded gasoline, now being phased out in the United States, is the main cause of lead poisoning.

Others metal air pollutants are compounds of mercury, arsenic, cadmium, and beryllium (coating on fluorescent tubes), mainly from the combustion of coal or from the smelting of ores.

Halogens (chlorine, bromine, fluorine, or iodine) are highly reactive and toxic in elemental form. Chlorine is widely used as a sterilizing agent in water treatment, producing small amounts of toxic chlorinated compounds that can escape into the air. Compounds containing both chlorine and fluorine, chlorofluorocarbons or CFCs, have been widely used as blowing agents for foamed plastics, as propellants for aerosol sprays, and as refrigerants (especially in car air conditioners).

Effect on the ozone layer

When they escape into the atmosphere, CFCs react with the ozone layer in the stratosphere in such a way that ozone is destroyed, while the CFC molecules are not. Consequently one CFC molecule can destroy many ozone molecules, creating what is known as a "hole" in the ozone layer that protects life on earth from damaging ultraviolet radiation. The production and use of CFCs is being reduced by international treaty.

Particulate matter is mainly in the form of *aerosols*, a suspension of solid or liquid droplets in a gas. These particles may be dust, ash, soot, lint, smoke, pollen, or spores. Natural sources contribute as much as 100 times those from human activities. Particulates are dangerous because they can reduce visibility, or get drawn into the lungs where they can damage tissue. The most dangerous particulates are cigarette smoke and asbestos, both of which can cause cancer.

Volatile organic compounds (VOCs) exist as a gas in air. Natural sources include isoprene and terpenes from plants such as evergreen trees (Blue Ridge Mountains), methane from swamps (will-o-wisp) and rice paddies, and from termites and ruminant animals by anaerobic decomposition. In the atmosphere these oxidize to carbon monoxide and carbon dioxide. Human activities add many other compounds, including such things as benzene, toluene, trichlorethylene, and formaldehyde (a preservative used in plywood glues).

Smog

Many of these pollutants are activated by sunlight and can react to form other compounds. *Photochemical smog* or brown smog results from these reactions, the color coming from nitrogen dioxide. Warm, dry climates create this condition usually in the late afternoon, in cities such as Los Angeles, Mexico City, and Lima, Peru. These smog conditions may be aggravated by a *temperature inversion*, where nighttime cooling (by radiation) traps cool air under warmer air, preventing dispersion of the smog and raising pollutant levels to those dangerous to human health, particularly for elderly and ill persons. By contrast, *gray smog* forms in winter and colder climates, from particulate matter ejected primarily from burning coal. In Denver, the burning of wood (urged by environmentalists) combined with a temperature inversion created dangerous conditions leading to a ban on wood-burning stoves.

The EPA under the Clean Air Act has power to regulate emissions, and several states and cities issue warnings when levels get too high.

Indoor air pollution

Indoor air pollution may be higher than outdoor, because of better insulation to conserve energy. Many home products contain chemicals for beneficial reasons, which can be released into the air within a home and rise to dangerous levels.

Radon, naturally occurring from decay of radium in rocks and soil, can accumulate in homes, mainly in the lower levels. The half-life of radon is only $3^{1}/_{2}$ days, but it decays into radioactive "daughters" with much longer half-lives. These are isotopes of polonium, lead, and bismuth, which can bind to dust and clothing, and if inhaled, to lung tissue. These daughters emit alpha, beta, and gamma radiation that are highly injurious to tissue.

Radon can also accumulate in groundwater and discharge into the air from showering, bathing, and toilet flushing. The EPA has established safe exposure limits to radon in homes. There are kits available to detect radon. Proper sealing of basements and adequate ventilation can reduce exposure to safe levels.

By far the most dangerous air pollutant is cigarette smoke. In the United States approximately 350,000 people a year die from diseases caused by smoking. Banning smoking would save more lives than almost any other pollution control measure. This is a controllable risk many of us illogically choose to accept.

Effects on humans

All of these air pollutants can cause bronchitis ("itis" means an infection), emphysema, asthma, fibrosis, and cancers in humans. They can cause acid rain, which kills vegetation, and this affects the habitat of living organisms at all levels. Acid rain damages not only habitats but buildings as well.

Controlling air pollution

Control of air pollution may involve many technological methods. Filters, cyclone collectors, and electrostatic precipitators can remove particulates. Fuels (both coal and oil) can be treated to remove sulfur or changed to low-sulfur sources, but at a cost. And the material removed still has to be disposed of. Fluidized-bed combustion reduces both sulfur and nitrogen oxide production. The "fluid" process, putting air through finely divided solids to create a "boiling liquid-like" surface, is used in catalytic reactions in oil refining and in wastewater treatment. This kind of reaction does not get rid of waste; it only changes it into a less dangerous and more easily controlled form.

Sulfur can be recovered chemically as elemental sulfur, or by scrubbing air exhausts with lime to form gypsum (calcium sulfate). Gypsum is used in making wallboard.

Hydrocarbon emission controls by catalysis can improve emissions from combustion by reducing them to water and carbon dioxide. A *catalyst* is a material that directs a chemical reaction without entering into it. Catalysis has long been used to produce useful chemicals in petrochemical plants. Modern automobiles all have a catalytic converter on their exhaust system, which uses platinum mesh as catalyst to convert carbon monoxide to carbon dioxide. Burning leaded fuel will "poison" the catalyst so it no longer does what it is designed to do; therefore, using leaded fuel in cars with such converters is illegal.

Volatile organics can be removed from the air by adsorption on activated charcoal. This process is used in dry cleaning plants and for treating air used to strip pollutants from groundwater and soil pollution. Spent charcoal can be burned, and sometimes reactivated, burning the sorbed materials as they are emitted during the reactivation process.

Air pollution standards

The 1970 Clean Air Act established primary standards to protect human health, and secondary standards to protect crops, climate, and personal comfort. Not only are not all goals reached, but continual political fighting goes on because of cost, with environmentalists on one side and industry on the other. The National Ambient Air Quality Standards (NAAQS) allow deviations, depending on location. National parks have the least permissible deviation; large cities the most, with up to 50 percent reduction in the standard to be met. A relatively new idea is *offset allowances*, which allow a site that is better than the requirements to swap or sell the excess amount with a site that does not meet the requirements.

Noise

Noise is a form of air pollution we generally do not think about or consider as such. The damage to hearing from noisy machinery both in factories and around the home (lawn mowers, leaf blowers, and similar equipment), rock concerts, boom boxes, car stereos played too loudly, is tremendous. Hearing loss is generally non-recoverable.

What you should remember

1. Air pollution is carried around the world by the winds and comes from both human and natural sources.

2. Point sources of air pollution are controlled by the Clean Air Act. Nonpoint or fugitive sources are difficult to control.

3. Smoking is a major cause of death from air pollution.

4. Acid rain comes from sulfur and nitrogen oxides emitted from burning fossil fuels, primarily from power plants.

5. Noise is a form of air pollution.

Chapter 18

Waste

Solid, Toxic, Hazardous

Introduction

Waste is something that somebody no longer wants or needs and hence wishes to discard. Since there is no "away," this means putting that waste someplace out of sight and out of mind, where it will not interfere with the daily living that produces still more waste. In earlier times, with far fewer people on this planet, this was relatively easy. There were lots of places to discard what was not wanted and plenty of new places to move to if one site became an intolerable place to live. Archaeologists and anthropologists delight in finding such disposal sites, because they provide a record of how people lived in such early times.

Today things are not so easy. Solid waste includes not only garbage and rubbish, but toxic and hazardous materials that are extremely dangerous to human life. The various methods of disposal, including incineration, land filling, reuse, and recycling, and methods of producing less waste included under the general heading of conservation, do not begin to solve the problem that as the world produces more and more people, the quantity of "waste" increases right along with the population.

The size of the problem

The EPA says we produce 11 billion tons of solid waste a year. That's about 40 tons, or 80,000 pounds per person per year. About half of this is agricultural waste, crop residues and animal manure. A lot of this gets plowed back as fertilizer, but this waste is the largest nonpoint source of water and air pollution in the country. Another third of solid waste is from mining operations, mostly stored near the point of production and, if properly handled, little cause for concern. Other industrial waste includes some 400 million metric tons (1 metric ton = 1000Kg = 2200 pounds) of which about a seventh, or 60 million metric tons, is toxic or *hazardous waste*. Municipal waste, garbage and refuse, is another 300 million metric tons, which amounts to about 2500 pounds per person per year. That is double what most other developed countries produce, and 5 to 10 times what the less-developed countries produce. What that says is that if you have a household of four people, you produce about 30 pounds of garbage and other nontoxic and nonhazardous refuse every day! A lot of this is packaging; a lot is newsprint.

Municipal refuse

Garbage is a monumental problem, especially in the large cities of the less-developed countries. Pollution and health problems are endemic in many places in the world because of the lack of proper recycling and disposal methods. People live on garbage dumps as a way of life to clothe and feed their families. In many of the affluent neighborhoods that have so-called housecleaning days, piles of discarded materials left for the waste collection services are picked over by passersby, as a sort of open-air, no-cost flea market.

Landfills

A properly designed landfill for municipal refuse is sited over impervious strata (and permits are harder and harder to come by), and constructed with impervious liners of compacted clay or synthetic materials. Every day the dumped material is covered with earth. When the landfill is finally full, it is covered with an impervious cap and planted.

Anaerobic (oxygen free) microbial action on organic matter (garbage) in such a covered landfill generates methane, which is a flammable gas. Collection piping is installed to prevent gas pressure buildup, and the methane is either flared off to reduce air pollution in older closed landfills, or it is collected and used as fuel for an engine driving a generator to produce electricity.

Monitoring wells and piping are installed to continually assure that no rainwater leaches out hazardous or undesirable materials into the groundwater. Such monitoring is required forever because no one knows how really secure such liners and caps will be in 10, 20, or 50 years.

Incineration

Incineration is a way of reducing the quantity of solid waste that must be handled, at the "cost" of producing large amounts of potential air pollution. Obtaining permits for the siting of an incinerator now can take years and millions of dollars. Even where there are technologically sound reasons for installing an incinerator, the NIMBY (Not In My Back Yard) syndrome raises costs, creates delays, and often puts insurmountable roadblocks in the way of incinerator progress.

Recycling and reuse

If we could separate all this material properly, much could be reused and recycled. Industrially there are a number of waste exchanges in the country, and more are being established. One big advantage of recycling is that it can save a tremendous amount of the energy required to produce new products, though often not as much as apparently possible because of the energy required in the collection, sorting, and distribution of recyclable materials. Costs are often high, but as nonrenewable resources become more scarce, recycling and reuse makes more and more economic as well as environmental sense.

Mandatory recycling

Voluntary recycling has not been entirely successful, because the organizations and people promoting these actions are only a small percentage of the largely indifferent population. It takes the mandate of law to initiate such action, and economic incentives to industries to install the necessary equipment, to really make recycling and reuse progress. Interestingly, in the industrial area many large companies that have installed corporate programs aimed at reduction of pollution have found that there are significant economic benefits in doing so, without tax incentives.

Hazardous waste

Hazardous waste is anything deemed detrimental to human health. Anything fatal to humans or laboratory animals at specific low dosages; or toxic, mutagenic, or carcinogenic to humans or other life forms; or that meets the definition of ignitable, flammable, combustible, toxic, corrosive, or explosive is considered to be hazardous waste. There are specific quantities that may be accumulated and shipped for hundreds of compounds, which must be reported if accidentally discharged into the atmosphere, water, or onto the ground. Everyone now handling such materials must follow cradle-to-grave recording and reporting of production, transport, and disposal of such materials. Because there is no uniform reporting standard (the federal government and various states have many different forms of records), anyone in that business today has a bureaucratic nightmare to contend with.

Illegal dumping

Illegal dumping of hazardous waste today, and indifferent dumping prior to current laws, has produced a multibillion-dollar problem in the United States. Many specialized companies have sprung up to contend with these problems. The initial small "mom and pop" companies have been supplanted by larger and larger specialists, as many of the smaller companies are bought up by the larger ones. One of the largest and the best known is Waste Management, headquartered in Oak Brook,

Illinois. Starting as garbage collectors many years ago, they are the most technologically advanced company in the world specializing in all phases of waste management.

Environmental laws

Congress had passed a number of regulations since the Environmental Policy Act of 1969, including the Toxic Substances Control Act (TSCA) in 1976, the Resource Conservation and Recovery Act (RCRA) the same year, and the Comprehensive Environmental Response Compensation and Liability Act (CERCLA), also known as Superfund. The Superfund Amendments and Reauthorization Act (SARA) was passed in 1990. It includes the right of communities to know the identity of hazardous materials to which they are exposed and provides that Material Safety Data Sheets (MSDSs) must be provided for all hazardous or potentially hazardous materials.

TSCA controls the discharge of toxic materials, RCRA the disposal of them, and CERCLA the cleanup of thousands of contaminated sites. Disposal methods for hazardous waste are similar to municipal waste, including incineration and landfilling, with much more stringent rules for both activities.

The EPA has expanded tremendously, is overburdened by problems, hammered by environmental activists and politicians, and facing a decades-long task, especially in Superfund activities. These problems will exist for years and hopefully will not overcome civilization for lack of a solution.

Infectious and radioactive waste

In addition to municipal and hazardous waste, there are two other categories, infectious waste and radioactive waste which are regulated. Infectious waste includes material from hospitals and medical and dental offices. Radioactive waste includes low-level radioactive material from nuclear medicine facilities, experimental laboratories, and nuclear power plants (tools, garments, and similar materials) and high-level radioactive waste, mostly from nuclear power plants and some experimental facilities. Each of these waste categories is handled differently in a manner primarily designed to protect humans.

What you should remember

1. We in the United States discard far more solid waste than any other country in the world.

2. Recycling and reuse of such materials in essential to reduce the problems of disposal and the consumption of nonrenewable resources.

3. There are four kinds of solid waste: municipal (garbage), hazardous, infectious, and radioactive, each handled differently.

Part 4

Our Planet's Future

Chapter 19

Urbanization

Introduction

Cities have existed for thousands of years. Ever since people first started congregating in family groups, those groups have gradually enlarged and ultimately developed into cities. Cities and civilization have developed side by side, dependent on each other. In cities, individual needs are supplied by a variety of sources, since the self-sufficient family group does not really exist in a city scenario. Cities have been centers, therefore, of not only the necessities of life, but also art, culture, education, and science. But along with those benefits urban living have come pollution, crowding, and disease.

Prior to the industrial revolution, most people lived in rural areas. As industries developed, more and more people moved into cities to answer the increasing need for industrial workers, until now nearly half the world's people live in there. It is projected that by the end of the Twenty-first Century between 80 and 90 percent of the world's population will live in cities. This transition is called urbanization.

Kinds of cities

There are different definitions of "city" depending on what country you are in. Generally a functional definition seems to work best. In a rural area residents depend directly on harvesting natural resources for their livelihood; in an urban area they do not. Our Census Bureau defines a Standard Metropolitan Statistical Area (SMSA) as having at least 100,000 people strongly tied to a central city of at least 50,000. A Consolidated Metropolitan Statistical Area (CMSA) has more than 1,000,000 people, formed by merging two or more population centers. Beyond 10,000,000 people, an area is considered to be a supercity, megacity, or megalopolis. There are many such megacities, formed by the overlapping influence of neighboring population centers, and industry plays an important role in influencing the character of the area. One such extends from Boston to Washington, DC. Another extends from Cleveland through Detroit, Chicago, Milwaukee, and down to St. Louis. Yet another extends from San Francisco to San Diego, taking in Los Angeles. Another is developing between Dallas and Houston.

Migration to the cities: United States

The United States has changed radically over two hundred years. In 1800, 95 percent of our population lived on farms or in small villages. Today, 75 percent of us live in cities, 23 percent in small towns (many commute to cities to work) and only 2 percent on farms. Part of that change has come from enormous improvements in farming efficiency. As fewer farmers were able to produce more food, some farmers migrated to cities to join the urban workforce. Chicago, for example, grew from 4000 people to a million in the fifty years between 1840 and 1890.

Migration to the cities: the world

The same city growth rate is true in the rest of the world, especially in the less-developed countries. In 1850, about 2 percent of the world's population lived in cities. By 1900, that had increased to 14 percent, and in 1988, nearly 42 percent were city dwellers. By 2000, more people will live in cities than in rural areas. The growth in third world countries has been and continues to be phenomenal. In 1900, London, with over 6.5 million people, was the largest city; New York was second; and Chicago was sixth, with 1.7 million. According to *World Population: Fundamentals of Growth* (Population Reference Bureau, Inc., Washington, DC, 1995), by 1950, New York was the largest city, London was second, and Tokyo third. Mexico City was not on the list of the top ten, nor was Chicago. By 1985, only 35 years later, Mexico City was the world's most populous city with 18 million people, and New York had dropped to fourth place. The estimates for 2015 are staggering: Tokyo will be the largest city with 29 million people! And Jakarta will be fifth with 21.2 million; it wasn't even on the top ten list fifteen years earlier. New York is estimated to be eleventh with 17.6 million. Twenty-seven cities will have populations of more than 10 million, seven of them more than 20 million.

Causes of urbanization

The causes of this escalating migration to cities are many, all influenced by the exploding population on earth. People are pushed into cities by dwindling food supplies, by local overpopulation, and by many political, religious, racial, and economic factors. People are also drawn to cities by a chance for jobs, excitement, freedom from village customs, and the expectation of enhanced prestige. Governments are located in cities, and because of this cities enjoy better communications, housing, job opportunities, and education, especially for those in power.

City problems

There are problems that attend cities' growth, particularly in less-developed countries. Traffic congestion and air pollution are endemic, as is lack of sanitation and concomitant disease. Housing is inadequate for most, slums are common, as are shanty towns and squatters' communities. The latter are popular but unauthorized, lack clean water, roads, sewers, and are often built on land unsafe for human habitation. The Union Carbide disaster in Bhopal, India, a few years ago is an example of what can occur when the urban poor settle in a mass near a large factory. Yet millions of people live, raise families, and educate their children in such circumstances. Crime rates increase dramatically also in these areas, and many cities are unsafe, especially for foreigners.

In the more-developed countries, overcrowding in cities is one cause of a rising crime rate and the deterioration of facilities such as housing and sewers. People fleeing these conditions by moving to suburban areas leave behind a progressively worsening situation because they take with them the money, education, and often the jobs people need. We read about this every day in the papers.

Impact of transportation

Improvements in transportation have probably had as much to do with urbanization as anything. Cities have always been located where transport is relatively easy—on rivers, crossroads, river fords, and centers of agriculture. Better modes of transport, including improved roads, have increased the ease of getting needed supplies and materials into cities, which stimulates city growth,

A typical example of early efforts to provide better transportation is the Illinois-Michigan Canal, which once provided a waterway from the Mississippi (via the Illinois River) to Lake Michigan, opening up Chicago to trade with St. Louis and the eastern and southern parts of the United States.

Groundbreaking for the canal was held on July 4, 1836, and the canal was completed and opened for traffic in 1845, even with a four-year hiatus caused by a depression. Despite competition from the railroads, the canal hauled freight on barges pulled by mules, and passengers on barges pulled by horses, for almost a century. The last passenger barge, the *Niagara,* came to Lockport, Illinois, in 1914, but freight traffic continued until 1935, when the larger Ship and Sanitary Canal was officially opened. The I&M Canal still exists as a recreational waterway, with headquarters in Lockport, Illinois, as a historical museum. The Ship and Sanitary Canal carries millions of tons of freight annually to and from Chicago to this day.

The building of freeways (or toll roads), following the example of the autobahns of Germany, has been a major factor in improving transportation. Most of us travel by car, and freeways opened up the country. Freeways have also caused many problems, however. They have wrecked neighborhoods, brought traffic noise and pollution, and stressed the capacity of cities to park and manage the flow of cars. One of the big problems we must solve is the transportation tangle, and one of the best ways to attack that is through increased use of improved public transportation.

What you should remember

1. There has been a steadily increasing migration from rural areas to cities starting with the industrial revolution, especially as mechanization of agriculture has reduced the need for large farm populations.

2. There has been an explosion in the growth of city populations in the less-developed countries. Tokyo, it is estimated, will contain nearly 30 million people after the end of this century.

3. Among the reasons for city growth are overpopulation in rural areas, poor food supply, religious, and racial problems, and lack of jobs in rural areas.

4. Improved transportation has had a major influence on city growth.

5. Crime, deteriorating housing, poor sanitation, and poverty are endemic to all large cities today.

Chapter 20

Planning
for
Sustainable
Cities

Introduction

The spread of populations to suburban areas is making cities out of them and, in the process, creating problems with traffic, zoning, and city services. This requires careful and detailed city planning. City planning has had a long history. It is a good career for today, involving the study of architecture, environment, pollution, traffic, land use, law, and many other disciplines.

Planned communities

Many "planned communities" have arisen in the United States, based at least in part on communal living, mostly religious in background, often with rigid rules of conduct and mores, particularly sexual ones. Most have failed, often because the sexual proclivities of the founder ultimately caused a breakdown in obedience to the rules! Some have lasted, such as the Amana Colony in Iowa. During the 60s "revolution," many communes arose as a backlash movement to the kind of world idealistic young people saw and did not like. Some of these have survived, but most have not.

Planned communities arose in many places in the world in this century, all trying to overcome the problems of land use, overcrowding, transportation, and psychological well-being. Many were successful for a time, because they brought low cost individual housing to moderate- and low-income families. Some of these structures were so "ticky-tacky" they have not lasted well. Some have converted into conventional suburbs. Some, like Park Forest, a suburb south of Chicago, started not so well but have survived.

Some towns, Boulder, Colorado, among them, have wonderful educational facilities and a strong environmental ethic, and have grown rapidly. Boulder remains successful because it rigidly controls expansion into the surrounding undeveloped area. Others, like the area around Disney World in Orlando, are still experiencing growth and slowly becoming as overcrowded as many other places.

Sun Belt problems

Movement of people to the Sun Belt states has posed tremendous problems in water supply, sewage treatment, and transportation. Nearly all rapidly growing areas have water supply problems of one kind or another. Expansion horizontally has created problems of land conversion, and expansion vertically creates other problems, such as excessive population density, air pollution, wind problems, and others.

One interesting but economically unsuccessful endeavor has been continuing for years north of Phoenix. A city called Arcosanti, designed by the architect Paolo Soleri, was to be self-sufficient, heated and cooled by solar and wind power, and it also incorporated a host of other interesting ideas. It is worth a visit if you are ever in that area. But it has not attracted the permanent residents Soleri had hoped for.

Successful city planning

Most designs call for a mix of parks, resting places, stores, homes, walkways, and a shopping mall that is a place to stroll, meet friends, or listen to a debate or music. Downtown Highland Park, Illinois, a suburban community of about 35,000 residents north of Chicago, was dying because of competition from Northbrook Court, a major nearby shopping mall. The mayor spearheaded a movement to rebuild the downtown area using tax-increment financing, and he succeeded over much opposition, mostly by residents who feared tax increases. The results have been wonderful, and a visit on a summer evening or a Saturday afternoon shows what can be accomplished with proper planning and a determined city government.

The most common way cities attempt to control the use of their land in a reasonable and productive fashion is through zoning. If you have a chance to attend a zoning hearing in your town, it is worth the time to see how the system works and how attempted changes in zoning can arouse the passions of neighbors.

A major problem all cities face is transportation. Private automobiles are used for about 95 percent of all trips and account for about 90 percent of all passenger miles. This is a tribute to the automobile industry, a source of tremendous profit to the petroleum industry, and a major energy, parking, and pollution problem for all of us. Mass transit and the replacement of fossil-fueled vehicles are essential for the future of livable cities.

Many countries and some cities have succeeded very well in controlling population, stabilizing city growth, and reversing the rural to city migration. Sri Lanka and China are good examples, showing it can be done. It is essential for the more-developed countries to contribute to the progress of the less-developed countries if the world is to ever gain a stable configuration.

What you should remember

1. The growth of suburban areas around large cities is bringing to them the same problems large cities have.

2. Good city planning is a necessity to maintain stable and sustainable population centers.

3. Private transportation accounts for about 90 percent of passenger miles today. The use of mass transit rather than private vehicles is an important requirement for city stability in the future.

Chapter 21

Ecology

Introduction

Nothing in nature exists in a vacuum. Everything alive (*biotic*) and not alive (*abiotic*) is intertwined with everything else. If there is a Law of Nature, similar to the Laws of Matter and Energy, it is that "You cannot change anything in nature without affecting many other things, more often than not in a totally unpredictable manner." This chapter presents a few examples.

Lamprey eels

Some years ago a species of eel was accidentally introduced into the Great Lakes. In time these lamprey eels greatly reduced the population of whitefish and lake trout by attaching themselves to the fish and eventually killing them. One result of this was that the population of alewives, a small herring-like fish that was a food for the whitefish and trout, multiplied at an enormous rate. Dead alewives piled up on the beaches and rotted; no one living nearby can forget the smell! Eventually the eels were brought under control by poisoning their breeding streams.

A program to control the alewives was inaugurated by importing coho and king salmon from the Pacific Ocean into the Great Lakes to, hopefully, use the alewives as food. That program succeeded very well, the salmon thrived at a rate far greater than expected, and a sport-fishing industry emerged as one result of the lamprey eel coming into the Great Lakes. Who could have predicted that?

Florida canal

In Florida the U.S. Army Corps of Engineers dug a canal to control flooding in the center of the state in an attempt to solve some of the problems confronting agriculture in that area. The canal diverted water from the Kissimmee River and has seriously affected the Everglades, which normally receives that water, with attendant problems for all the wildlife there. Changes in water flow in the river have allowed a different variety of tree to establish itself along the riverbanks. Leaves from those trees affect the riverbed when they fall into the water, with disruption of the fish and animal life dependent on the river for food and water.

In some areas there has been rampant growth of water lilies, to such an extent that they make the waterway unusable for small craft navigation. Removal of the plants by dredging has created shoreline problems where they are piled, so efforts were made to burn them. The resulting ash was washed into the water by rainfall, and being rich in nutrients, merely provided food for many other undesirable varieties of plants. "Undesirable" in this instance is a human definition, not one of nature's.

Recently there have been some experiments to redirect part of the water back into the Kissimmee River. Over short sections, perhaps 10 to 15 miles long, the increased water flow is starting to bring about a return to the conditions that existed before the diversion canal was constructed.

DDT

Another example of the interrelatedness of living things can be seen in the use of DDT to control insect pests. Initially such control was seen as a boon to agriculture. Gradually it became evident that there were unforeseen results. DDT dust carried by the wind fell upon the surface of lakes, rivers, and the oceans. The dust was absorbed by organisms at the bottom of the food chain, and the chemical gradually working its way up the chain until it affected the strength of the shells of eggs laid by many species of birds. The end result of using DDT to protect food crops from insect depredation was to seriously affect the ability of some bird species, many far removed from the sites where DDT was used, to survive.

Zebra mussels

Recently the Great Lakes have suffered an infestation of a small shellfish called a zebra mussel. These were apparently carried in the ballast tanks of ocean-going vessels that entered the Great Lakes via the St. Lawrence Seaway. With no natural predators, the zebra mussels proliferated exactly where they could cause real problems, along water intakes for drinking water and cooling water for power plants.

Efforts to control the population of these shellfish are hampered by the realization that doing one thing to reduce the mussel population may create other, perhaps unrecognized problems. Meanwhile some scientists have discovered that the zebra mussel may be a source for a cancer-inhibiting drug. Also, the mussels strain particulate matter from the water, increasing the amount of sunlight that reaches the lake bottom, thus promoting the growth of algae. This affects the food chain and, as the algae die and decompose, it also affects the taste of potable water taken from Lake Michigan.

Yellowstone fire

One of the most striking illustrations of the complicated interrelationships in the environment occurred in Yellowstone National Park during the summer of 1988. That summer was unusually dry, and the meadows were parched and brown. There had been no major fires for years, and the forests were tinderboxes of living and dead timber waiting for something to ignite them. Some man-made fires started, but most occurred, as they do every year, from lightning strikes. In 1988 the combination of circumstances created firestorms that burned hundreds of thousands of acres of forest. High winds carried embers far and wide, and the fires were so fierce that they created their own winds to spread the flames.

The policy of the Park Service had been to let small fires burn themselves out unless human life or property were endangered. Under pressure from the public, the media, and eventually the federal government, that policy changed. Everyone felt that Yellowstone was going to be destroyed forever. So a massive and costly effort was mounted, involving thousands of people, to put the fires out, all to little avail. Finally, after more than three months, nature extinguished the blazes with snowfall.

It took a while to realize that such events had occurred many times during the thousands of years Yellowstone has existed and that just as rain is essential for a tropical rain forest, so is fire essential for the survival of Yellowstone. Accumulations of undergrowth and dead timber that were stifling the regrowth of some tree species were removed by the fire. The resulting ash replaced much needed nutrients into the soil and rivers as snow melted and as rain began in spring. Small burrowing animals that survived the fire provided easily accessible food, with the meadows and underbrush cleared away, for predator birds and animals. Larger animals such as deer, elk, moose, and bison were little affected by the fire itself; they merely moved out of the way. But the loss of easily available food stressed them greatly during the harsh winter. Only the strongest survived; the weak died. And in their dying they provided a bonanza of food for other predators, whose numbers had been diminished because of the lack of available food.

Spring brought an outburst of growth, due to the increased nutrients and sunlight in the fire-thinned forest. Two years later, Yellowstone, though showing obvious signs of the fire, was bursting with new growth and life. And scientists have had an unparalleled opportunity to study at first-hand part of the ever-occurring long-term cycle of life and death that has been Yellowstone's history through the centuries.

What you should remember

1. Everything in the environment is interrelated.

2. You cannot change any one thing without affecting many others.

3. It is difficult, perhaps impossible, to predict what the end results of changes will be.

Chapter 22

Looking
Ahead

Introduction

Where does the world go from here? There are as many ideas and positions as choices about how we must manage our populations, resources, food supplies, air, water, and land. What seems clear is that we are on a course leading to increased difficulties and problems, if not downright disaster, unless we make radical changes in the way we humans live on and manage this earth we all inhabit. There is much evidence to suggest that we have gone beyond the point of recovery, and much to suggest that human ingenuity can overcome our problems if—and it's a big *if*—we can change attitudes.

Gaia

James Lovelock, a British scientist, propounded a theory that the earth is a living entity, which he calls *Gaia*. His idea is that conditions here on our planet are unique in the solar system, and that everything on earth interacts in such a way as to maintain those conditions. There is an excellent computer game called *SIMEARTH* that allows the player to vary conditions in an attempt to develop a stable civilization on earth. The program develops a life of its own, and if you finally get a stable earth established, changes that occur are reacted to in such a way as to maintain that stability. Interestingly enough, Lovelock's idea is gaining respectability in the scientific world.

Need for population control

One thing appears certain, that somehow control of population growth is the key to attaining a sustainable earth, one in which the human race can live within the carrying capacity of the planet. And that means a more equitable distribution of the earth's resources. Hungry people have little use for philosophy. So we need a goodly application of what we usually think of as justice, sort of a universal application of the principle "to each according to his needs, from each according to his ability"; a universal exchange of scientific information; an agreement on what constitutes ethical principles; a good dose of hope; and enough self-interest to make it all work.

Environmental ethics

Ethics is the branch of philosophy concerned with morals (difference between right and wrong) and values (the worth of actions or things). Environmental ethics asks questions and attempts to answer them concerning the relationship of humans to the natural environment. These questions are not new. Anyone reading the so-called *GREAT BOOKS* discussed in the classes started by Mortimer Adler at the University of Chicago has had to grapple with these ideas.

A scientist, Paul Taylor, has set forth three basic principals of ethical conduct. Paraphrased, they are as follows:

❑ Do not do any harm to a natural entity.

❑ Do not interfere with natural systems.

❑ Do not mislead any organisms capable of being misled.

Those ideas led Taylor to vegetarianism, to an end of hunting and fishing, and to the protection of wilderness areas. Moral rights, however, do not extend far from humans, and not at all to nonliving natural objects. This is in conflict with Lovelock's concept of Gaia, the earth as a living entity.

Stewardship

One response to this is the idea of stewardship, which says that humans have a responsibility to care for the world. Many companies today have adopted this idea with regard to the impact of their products on the environment, with a cradle-to-grave oversight of what they sell, how it is used, and what happens to it.

Environmental laws

A growing awareness of stewardship has led to a multitude of environmental laws, not only in this country but also in many of the European countries. The areas in the world that were formerly the Soviet Union and the Eastern Block are suffering grave problems because they did not adopt this attitude in the past.

There is no real international body with authority to enforce international environmental laws. What happens is that governments sign treaties regarding such things as whaling rights, oil discharges, and air pollution, but if the countries break those treaties there are only moral sanctions to try to enforce them. Application of such sanctions is enmeshed in politics, so that even with the cooperative scientific work in Antarctica, for example, pollution there is growing because no one enforces the rules regarding waste.

Environmental organizations

There are literally dozens of environmental organizations concerned with one facet of environmental issues or another. As they grow from small, grass-roots organizations staffed by enthusiastic, idealistic people, they gradually get concerned with maintenance of membership to insure income, publication of slick magazines, lobbying efforts in Washington, and generally are very much in competition with each other.

Many of them do much good. The Nature Conservancy is the largest private owner of nature sanctuaries in the world. Even so, much of their time and money is devoted to maintaining the jobs and prestige they have acquired. Many of the smaller organizations have neither the money nor the patience to compete with the larger ones, and so they resort to violence and other aggressive behavior to try to attain their ends and gain publicity.

Lifestyle choices

All of this brings me finally to the fact that we each have to develop a personal philosophy and lifestyle in our own relationship with the environment in which we live. This list of personal activities, taken from Cunningham and Saigo's textbook, *Environmental Science*, is a good one to consider.

- ❑ Think globally and act locally.
- ❑ Continue your education. Read, study, discuss, take courses.
- ❑ Vote.
- ❑ Learn about your immediate area.
- ❑ Think about the consequences of your lifestyle and jobs.
- ❑ Work with others. Get active in groups that share your interests.
- ❑ Do not carry the world on your shoulders.
- ❑ Live simply and frugally.
- ❑ Do not get discouraged.

Edward Abbey was a reporter and author with a strong environmental interest. This is a quotation from one of his works:

> One final paragraph of advice; do not burn yourself out. Be as I am—a reluctant enthusiast, a part time crusader, a halfhearted fanatic. Save the other half of yourselves and your lives for pleasure and adventure. It is not enough to fight for the land; it is even more important to enjoy it. So get out

there and hunt and fish and mess around with your friends, ramble out yonder and explore the forests, encounter the grizz, climb mountains, gad the peaks, run the rivers, breathe deep of that yet sweet and lucid air, sit quietly for awhile and contemplate the precious stillness, that lovely, mysterious and awesome space. Enjoy yourself, keep your brain in your head and your head firmly attached to the body, the body active and alive, and I promise you this much: I promise you this one sweet victory over our enemies, over those desk-bound people with their hearts in a safe deposit box and their eyes hypnotized by desk calculators. I promise you this: you will outlive the bastards.

What you should remember

1. Population control is essential to a sustainable earth, as is an equitable distribution of the earth's resources.

2. There are laws in many countries and many treaties between nations designed to protect the world's environment. But there is no international agency with the power to enforce environmental laws worldwide.

3. Take time to enjoy the good things in life. You'll live longer!

To Learn More

There is a plethora of books, articles, magazines, videos, slides, and essays concerned with the environment. Surfing the Internet can add even more reading. Many viewpoints and special "agendas" are involved, creating some degree of confusion in understanding how the world works. The references below are a few that the author has found both interesting and educational. It is by no means a complete list, but it is a starting point for readers who want to expand their knowledge beyond this book..

Textbooks

The following four textbooks are college level, each with a slightly different approach to the subject, but they cover the subject matter in far greater detail than this book. They are usually not found in public libraries. Phone calls to community college and university bookstores in your area should turn up one or more that may be purchased.

Environmental Science by Cunningham and Saigo, Wm. C. Brown, publisher.

Environmental Science by Nebel and Wright, Prentice Hall.

Environmental Science by Miller, Wadsworth Publishing Company.

Environmental Science by Enger and Smith, McGraw-Hill.

Biosphere 2000 by Kauffman and Franz, Harper-Collins.

Internet Information

Environmental Guide to the Internet by Briggs-Erickson and Murphy, Government Institutes, publisher.

From the bookshelf

Some of the following may be available in public libraries; some may be available at bookstores.

> *The ABCs of Environmental Regulation* by Joel B. Goldsteen, Government Institutes. A plain-English introduction to the laws that protect our environment.

> *Biosphere 2* by John Allen, Viking Press. The story of the planning and construction of a totally closed ecosystem and the personnel that lived there for two years. A follow-up article detailing the problems that arose can be found in *Science News, The Weekly Magazine of Science* for November 16, 1996.

> *Gaia, An Atlas of Planet Management*, edited by Dr. Norman Myers, Anchor Books. "Our place on the planet and what we are doing to ourselves."

> *Readings in Risk,* edited by Glickman and Gough. Resources for the Future. Essays on various regulatory, technological, and health risk assessment.

> *The Dose Makes the Poison* by M. Alice Ottoboni, Vincenti Books. A plain-language guide to toxicology.

> *The Bug Book* by Barbara Pleasant, Storey Publishing. The life cycle of insects and safe insect control.

> *Agenda 21,* edited by Daniel Sitarz, Earthpress. Planet management-planning for a sustainable future.

> *Taking Sides,* edited by Theodore Goldfarb, Dushkin Publishing Group. Subtitled "Clashing Views on Environmental Issues."

> *Watersheds* by Newton and Dillingham, Wadsworth Publishing Company. Environmental ethics cases.

Annual compilations of essays

> *State of the World,* issued by the Worldwatch Institute, W.W. Norton & Company. An annual report on progress towards a sustainable society.

> *Annual Editions, Environment,* edited by John Allen, Dushkin Publishing Group. One of a series of annual collections of essays on various subjects.

Videos

There are numerous interesting and informative videos, many available at public or college libraries. Some are difficult to locate but worth the effort.

After the Warming. A two-hour NOVA video looking back at the world from the year 2050, after global warming has resulted in drastic changes in our planet, and the resultant formation of a Planet Management Authority.

A Desert Place. A 30-minute NOVA video about the Sonoran Desert ecosystem, its climate and life.

Out Of The Rock. A 30-minute video produced by the U.S. Bureau of Mines about the production and uses of minerals.

The Unfinished Song. A 20-minute video produced by Landis-Trailwood Films of Huron, SD, about the great Yellowstone fire and its effect on that ecosystem.

Tomorrow's Energy Today. Two videos, together totaling 49 minutes, produced by the U.S. Department of Energy.

Glossary

abiotic: Nonliving.

acid rain: Highly acidic rain caused by sulfur dioxide and nitrogen oxides in the air, primarily from burning of coal in power plants.

active solar heating: Using the sun's energy directly to heat water to produce steam to drive a turbine to turn a generator to produce electricity.

aerosol: A suspension of fine particles of liquid or solid in the air.

albedo: A measure of the ability of surfaces to reflect sunlight. The higher the albedo, the greater the reflection.

aquatic ecosystem: an ecosystem that exists in the hydrosphere, the watery part of our planet.

aquifer: An underground source of water.

artesian well: A self-flowing well caused by a tilted aquifer, the high portion of which is higher than the well at the surface.

atmosphere: The band of air surrounding the earth.

biomass: Mass consisting of living or once-living matter.

biome: An ecosystem on land.

biotic potential: The reproductive rate of a species.

biotic: Living or alive.

biosphere: The sphere of life, the earth, comprising the atmosphere (air), hydrosphere (water) and lithosphere (land).

boreal forest: A forest in a polar area, often called a **taiga**.

carbon cycle: *See* **photosynthesis**.

carcinogen: Cancer causing material.

carnivore: A meat eater

carrying capacity: The maximum population of a species an ecosystem can sustain indefinitely.

catalyst: A substance that promotes a chemical reaction without being part of that reaction.

chain reaction: A self sustaining radioactive decay process.

chaos: The random state of matter.

climate: Weather over a long period of time in a particular place.

cogeneration: Utilizing waste heat from burning fuel to drive an engine for other purposes, such as heating buildings.

community: A group of populations of different species existing together in a particular habitat.

Conservation of Energy, Law of: Energy can neither be created nor destroyed (First Law of Thermodynamics).

Conservation of Matter, Law of: Matter can neither be created nor destroyed, only changed in form.

cornucopians: Those who believe the world's resources can support unlimited growth.

conventional energy sources: Coal, gas, and oil.

critical air pollutants: The seven specified pollutants: sulfur dioxide, nitrogen oxides, particulates, hydrocarbons, photochemical oxidants, carbon monoxide, and lead.

deciduous trees: Trees that drop their leaves.

decomposer: Mainly fungi and bacteria, that break down dead organisms.

deserts: Biomes receiving less than 4 inches of water per year.

desertification: Turning productive land into wasteland from overgrazing, poor farming practices and deforestation.

drip irrigation: A system of pipes or hoses carrying water, with holes at the exact spot water is required.

ecology: The study of the interactions between organisms, living and nonliving, within environmental systems.

ecosphere: Everything within the biosphere.

ecosystem: An area, large or small, including all the living and nonliving entities therein.

electron: A light charged particle "revolving" around the nucleus of an atom.

energy: The capacity to do work.

entropy: Energy always runs "downhill," from a more organized state to a lesser one. The measure of such disorganization is known as entropy. The entropy of the world is always increasing, and the replenishment to get that energy back "uphill" comes from the sun.

environment: The total of all the external factors that affect the organism within a specific area.

erosion: The wearing away of topsoil by wind and water.

evaporation: The process by which matter changes from a liquid to a gas, by the input of energy in the form of heat.

external cost: Costs, such as damage to the environment, other than what a consumer pays for a product.

First Law of Nature: You can not change one thing without affecting many others.

First Law of Thermodynamics: *See* **Conservation of Energy, Law of**.

food web: Complex mixture of herbivores, carnivores, and omnivores.

forest: A land area with sufficient water supply to support the growth of trees.

fugitive air pollutants: Materials in the air the source of which is difficult or impossible to locate.

Gaia: James Lovelock's idea that the earth is a living entity, which reacts to maintain its own good health (from *Gaia*, Greek goddess of the earth).

gasohol: Automotive fuel containing a mixture of gasoline and alcohol.

genetic assimilation: Gradual evolution that changes one species into another different one.

gray smog: Formed in winter by particulate matter ejected from smoke stacks and chimneys.

green manure: Crops planted and then plowed back into the earth to decompose and add nutrients to the soil.

greenhouse effect: Warming of the earth because atmospheric composition interferes with radiation of heat back into space.

green revolution: High yield crops developed to improve food production, especially in the underdeveloped countries.

groundwater: Water that exists in saturated ground beneath the zones through which surface water percolates. The top of this saturated layer is known as the water table.

habitat: The particular place and set of conditions that favors a particular species (plant or animal).

half-life: The rate of decay of a radioactive element, specific to each element, that measures the time for half of the element to be used up.

hazardous material: Any material either specifically listed by the U.S. EPA as hazardous, or exhibiting the characteristics of being explosive, combustible, toxic, or corrosive.

herbivore: A first trophic level consumer that feeds on plants.

hydrologic cycle: The process whereby sunlight energy evaporates water (mostly from the oceans), which rises and condenses in the colder upper atmosphere, releasing the heat of evaporation and creating clouds and rain, which falls to earth to repeat the cycle.

hydrosphere: That portion of the world (the biosphere) that exists in water.

igneous rocks: Rocks crystallized out from the molten mantle in the interior of the earth.

internal cost: The cost that a consumer pays for a product, generally covering manufacturing raw materials, labor, and profit. It does not include the environmental costs of producing that product.

ionosphere: That portion of the earth's atmosphere where the molecular composition is so rarified that molecules can be ionized by radiation from the sun.

isotopes: Forms of elements with the same chemical properties but different molecular weights because of the presence of neutrons in the nucleus of the atom.

lithosphere: That portion of the biosphere consisting of the earth's crust.

malnutrition: Not receiving sufficient protein, vitamins, and other necessities in ones diet, even though there may be enough carbohydrates.

Malthusian growth: A growth pattern described by Thomas Malthus that consists of rapid exponential growth followed by a population crash as food supply is depleted.

mass: The scientific name for matter, anything that takes up space and has weight.

mesosphere: That portion of the atmosphere above the stratosphere.

metamorphic rocks: Igneous and sedimentary rocks changed by heat and pressure.

meteorology: The study of weather and climate.

monoculture: Planting of one species of crop only.

neomalthusians: Believers in the modern world in the theory of Thomas Malthus about population growth and crash.

neutron: A heavy particle in the nucleus of an atom that does not carry an electrical charge.

NIMBY: "Not in my back yard."

nonpoint sources: Sources of air or water pollution that are hard to define, such as odors or storm-water runoff.

nonrenewable resources: Those that cannot be renewed within the human time frame.

offset allowances: A system of allowing pollution above permissible limits to be balanced against pollution below those limits.

omnivore: A meat and vegetation eater.

organisms: Living things.

overshoot: The extent to which a population growth exceeds the carrying capacity of its environment.

ozone layer: A portion of the stratosphere containing ultraviolet absorbing ozone.

passive solar heating: Positioning buildings to take advantage of the sun's heat.

pest: Anything that interferes with the smooth flow of human life.

photochemical smog: Air pollution caused by chemical reactions between various pollutants, stimulated by sunlight.

photosynthesis: The process by which plants combine carbon dioxide and water to form complex carbohydrates and emit oxygen under the influence of sunlight energy.

photovoltaic cells: Devices that convert light energy directly into electricity.

point sources: Well-defined sources of air and water pollution, such as a smokestack or a sewer outfall.

population: All the members of a specific species occupying a specific place.

productivity: The amount of biomass produced in a specific environment in a specific period of time.

power: The rate of expenditure of energy.

primary air pollutant: Materials entering the air directly in harmful quantities.

primary consumer: An organism that feeds on vegetation.

primary producers: Plants.

proton: The heavy charged mass in the nucleus of an atom.

radioactive decay: The spontaneous breakdown of radioactive elements into other elements or isotopes, with the emission of one or more types of radiation. *See* **half-life**.

radioactivity: The property of an element that causes it to spontaneously decay into an isotope of the same element or another element entirely, giving off radiation in the process.

radon: A radioactive gas formed from the decay of uranium.

renewable resources: Resources that can be replaced within the human time frame. Plants and fresh water (from rain) are examples. All depend on energy from the sun, a perpetual resource.

respiration: The reverse process of photosynthesis.

savanna: A biome receiving between 4 and 10 inches of rain a year, thus supporting grasses and shrubbery.

scavenger: An organism that feeds on dead organisms.

secondary air pollutant: Harmful materials created by chemical or physical changes after they have entered the air.

secondary consumer: An organism that feeds on primary consumers.

sedimentary rocks: Rocks formed from sediments in the oceans, past and present, compacted by pressure.

stratosphere: The portion of the atmosphere from about 30,000 to 100,000 feet above the earth.

sublimation: The process whereby a solid turns directly into a gas without going through the liquid phase.

synthesis: The production of a substance by the union of chemical elements, groups, or compounds.

taiga: A forest in a polar area. *See* **boreal forest**.

temperature inversion: A condition in which a layer of warm air forms above a layer of cold air at the earth's surface and traps pollutants.

terrestrial ecosystem: an ecosystem that exists in the lithosphere, the land part of our planet. *See* **biome**.

tertiary consumer: An organism that feeds on secondary consumers.

thermodynamics: The study of the movement of energy.

thermosphere: The layer of atmosphere above the mesosphere.

trophic levels: Consumer levels, from the Greek word for food.

troposphere: The layer of atmosphere from the surface of the earth up to the stratosphere.

tundra: Polar savannah, or grassland in the cold north.

urbanization: The movement of populations into cities from rural areas.

weather: The short-term description of atmospheric conditions in a particular place at a particular time.

Index

A

B

C

GOVERNMENT INSTITUTES
MINI-CATALOG

PC #	ENVIRONMENTAL TITLES	Pub Date	Price
585	Book of Lists for Regulated Hazardous Substances, 8th Edition	1997	$79
4088	CFR Chemical Lists on CD ROM, 1997 Edition	1997	$125
4089	Chemical Data for Workplace Sampling & Analysis, Single User	1997	$125
512	Clean Water Handbook, 2nd Edition	1996	$89
581	EH&S Auditing Made Easy	1997	$79
587	E H & S CFR Training Requirements, 3rd Edition	1997	$89
4082	EMMI-Envl Monitoring Methods Index for Windows-Network	1997	$537
4082	EMMI-Envl Monitoring Methods Index for Windows-Single User	1997	$179
525	Environmental Audits, 7th Edition	1996	$79
548	Environmental Engineering and Science: An Introduction	1997	$79
578	Environmental Guide to the Internet, 3rd Edition	1997	$59
560	Environmental Law Handbook, 14th Edition	1997	$79
353	Environmental Regulatory Glossary, 6th Edition	1993	$79
625	Environmental Statutes, 1998 Edition	1998	$69
4098	Environmental Statutes Book/Disk Package, 1998 Edition	1997	$208
4994	Environmental Statutes on Disk for Windows-Network	1997	$405
4994	Environmental Statutes on Disk for Windows-Single User	1997	$139
570	Environmentalism at the Crossroads	1995	$39
536	ESAs Made Easy	1996	$59
515	Industrial Environmental Management: A Practical Approach	1996	$79
4078	IRIS Database-Network	1997	$1,485
4078	IRIS Database-Single User	1997	$495
510	ISO 14000: Understanding Environmental Standards	1996	$69
551	ISO 14001: An Executive Repoert	1996	$55
518	Lead Regulation Handbook	1996	$79
478	Principles of EH&S Management	1995	$69
554	Property Rights: Understanding Government Takings	1997	$79
582	Recycling & Waste Mgmt Guide to the Internet	1997	$49
603	Superfund Manual, 6th Edition	1997	$115
566	TSCA Handbook, 3rd Edition	1997	$95
534	Wetland Mitigation: Mitigation Banking and Other Strategies	1997	$75

PC #	SAFETY AND HEALTH TITLES	Pub Date	Price
547	Construction Safety Handbook	1996	$79
553	Cumulative Trauma Disorders	1997	$59
559	Forklift Safety	1997	$65
539	Fundamentals of Occupational Safety & Health	1996	$49
535	Making Sense of OSHA Compliance	1997	$59
563	Managing Change for Safety and Health Professionals	1997	$59
589	Managing Fatigue in Transportation, *ATA Conference*	1997	$75
4086	OSHA Technical Manual, Electronic Edition	1997	$99
598	Project Mgmt for E H & S Professionals	1997	$59
552	Safety & Health in Agriculture, Forestry and Fisheries	1997	$125
613	Safety & Health on the Internet, 2nd Edition	1998	$49
597	Safety Is A People Business	1997	$49
463	Safety Made Easy	1995	$49
590	Your Company Safety and Health Manual	1997	$79

Electronic Product available on CD-ROM or Floppy Disk

PLEASE CALL OUR CUSTOMER SERVICE DEPARTMENT AT
(301) 921-2323 FOR A FREE PUBLICATIONS CATALOG.

Government Institutes
4 Research Place, Suite 200 • Rockville, MD 20850-3226
Tel. (301) 921-2323 • FAX (301) 921-0264
E mail: giinfo@govinst.com • Internet: http://www.govinst.com